War or Wealth

How Technology Brings War With China to the Heartland - and Decides America's Future

Bert Kastel

How to Consider Technology
When Confronting China?

This book gives the answer.

https://WarOrWealth.com

* * *

Published in the United States by Competitive Edge International, Inc. of Sarasota, Florida (USA).

Paperback ISBN 979-8-9868573-0-5
Hardcover ISBN 979-8-9868573-2-9
eBook ISBN 979-8-9868573-1-2

Table of Contents

"The consequences of a war between China and the United States would be catastrophic for humanity.

If a war were to break out between them, it would probably be the most important war in human history. It would dwarf all previous wars in scale and would have enormous consequences for the future of humanity.

War between the United States and China would not only cause massive loss of life but would also lead to a global economic collapse.

The conflict would be so devastating that it could take decades for the world to recover.

If a war were to break out, it would likely be fought over Taiwan."

Galileo AI, May 26, 2022
A state-of-the-art Artificial Intelligence called *Galileo* answering the question:
"What would be the likely consequence of a war between
the United States and China for the future of humanity?"

6 Crucial Facts About China

Chinese History in 3 Minutes

"We will never allow anyone, any organization, or any political party, at any time or in any form, to separate any part of Chinese territory from China. ... We do not promise to renounce the use of force and reserve the option to use all necessary measures."

Xi Jinping, October 18, 2007,
at the Chinese Communist Party's (CCP)
19th Party Congress[1]

To get our readers on a similar starting page, I have put together a list of six important facts about China's history. In public debate, they often are unknown or not sufficiently considered.

1. **China was the biggest economy on Earth until it was passed by the United States at around 1900.**
   For thousands of years of human history, China and India had been the two most populous world regions, usually with the biggest economies on Earth.[2]
2. **European powers aggressively carved out colonial outposts in parts of China after 1839.**
   These included control over Hong Kong, Macao, Shanghai, and "spheres of influence" over some portions of the Chinese Empire. During that time, the creation of a Christian Heavenly Kingdom in Southern China resulted in the bloodiest civil war of human history, killing about 20 million people in the early 1860s.

3. **Japan conquered the most important parts of China in the 1930s and 1940s in a brutal military campaign resulting in over 20 million casualties.**
   Japan's conquests had started in the 1890s and focused on China's coastal population centers. During World War II, and aided by U.S. economic and military aid, the Chinese Nationalists and Communists together ousted the Japanese while the U.S. drove the Japanese military to retreat from Southeast Asia and the Pacific.

4. **The Kuomintang under Sun Yat-sen overthrew the emperor in 1911 and established the Republic of China (ROC).**
   The ROC's rule was constantly challenged by Communist forces. In 1949, they had to retreat to what then became Taiwan. There, they established authoritarian rule while still considering themselves the rulers of all of China.

5. **Until China's opening to the U.S. and the West in the 1970s, the ROC in *Taiwan* held China's permanent seat on the U.N. security council until 1971, representing all of China.**
   Until the late 1980s, the Kuomintang government of Taiwan pursued unification with the mainland as a political goal. Taiwan's Western-style democracy started taking hold in the beginning of the 1990s.

6. **Starting at around 1980, China under Deng Xiaoping introduced capitalist forces into China, catapulting one of the poorest countries on Earth to become the world's biggest economy again.**
   This was the most rapid and largest economic development of any large country in human history, with average growth rates exceeding 10% per year for about four decades.

China considers recent Taiwanese reforms a temporary blip and pursues the re-establishment of Chinese boundaries as they existed before Europeans established footholds in the 1830s. It continues to hold the position previously shared with Taiwan that the ROC and the mainland are part of one nation.

Preface

Delivering Peace and Prosperity Using 21st Century Tools

Geography matters - but technology even more. One generation ago, the Internet had just become "a thing." Two generations ago, even mainframe computers barely affected anybody's life. And it was *three* generations ago, long before personal computers existed, that we established the instruments and institutions still dominating the global system. We called them the "rules-based liberal world order." Or simply: the free world.

It worked exceptionally well, for us and for most of humanity, by almost every large-scale measurement.

Today, three generations on, geography still is important. The oceans, mountains, and islands still exist. But that old world and its order are no more. Today we have sensors documenting and artificial intelligence analyzing nearly everything we do, almost everywhere. Devices can connect any person, thing, and place in the world instantaneously. We all now own and use supercomputers masked as phones. Each one of these is at least a million times more capable than what sent men to the moon 50 years ago. Computing power gave us just-in-time supply chains, gene sequencing, satellite mega-constellations, and self-driving electric cars.

But even this is old news. The pace of digitization and technological disruptions is not slowing down. It keeps growing at exponential speeds. Computing capacity still doubles every two years. Some changes may even *accelerate*. In the early 2020s, artificial intelligence algorithms have been increasing their abilities *tenfold each year*.[3] It is equivalent to improving by over one billion percent in a decade. All this keeps transforming every aspect of our lives.

Except, it seems, our global "international" system. We still call it rules-based, liberal, and free. We still rely on the same tools to protect it. And we still use the same institutions and approaches to foreign affairs. It is as if we could ignore the impact of technological disruptions.

But we can't. Technology transforms everything around us, in plain sight. The world shifts and most of us don't know how to adjust. Many then simply cruise along, focusing on the gadgets and ignoring their deeper impact.

My mission is to help people make sense of such changes, and better prepare them and society for the future. I want this future to be prosperous and peaceful. I want people to be empowered. I want to live in a free world.

But we cannot have either if we ignore reality. More specifically, we can neither have security, prosperity, or opportunity if the great powers fight each other.

War or wealth. That is the decision we have to make.

We have been here before, 100 years ago. It did not go well then. Today, though, we still have a chance.

But I am getting ahead of myself.

In the fall of 2021, I organized an online "flash symposium" to assess the geopolitical impact of the U.S. withdrawal from Afghanistan. It involved a series of a dozen conversations with representatives of reputable think tanks from twelve countries: Japan, the Philippines, Australia, India, Pakistan, Ethiopia, the UK, France, Italy, Germany, Estonia, and the Ukraine. One contribution of this diverse group of thinkers affected me more than others.

The conversation with Hugh White from the Strategic and Defence Studies Centre (SDSC) at the Australian National University inspired me to write this book. He quickly jumped into what seems to be the major foreign policy challenge of our generation: the rivalry between the United States and China. Hugh phrased his arguments in a compelling, no-nonsense way.

Others added different perspective. For example, Tetsuo Kotani from the Japan Institute of International Affairs (JIIA) pointed to the urgency of China's challenge in Asia. He said that Japan would have to be prepared to fight alongside the U.S. to defend Taiwan. Mario Del Pero from the Italian Institute for International Political Studies (ISPI), on the other end, stressed the need to trust the "long arc of history" even in geopolitics.

Hailing from Australia, it was Hugh White who seemed at the same time close and far enough from the issue to bring it all together best. He pointed out the obvious hard power shifts that already have directly impacted all countries in the Asia-Pacific and beyond. Then, he stressed how this means that not just Australia but above all the U.S. must make some hard choices. And he is right.

Hugh is also in the best company when saying so. For example, John Mearsheimer, Kishore Mahbubani, Graham Allison, and Henry Kissinger also keep asking to acknowledge new realities and shift our views and strategies on global affairs. Many of their arguments point at how underlying trends have resulted in a changed global landscape, one that continues evolving.

However, all are still too tame about the contribution of the most significant driver of this change: **technology**. It is not just about artificial intelligence (AI), as Allison, Kissinger, and Eric Schmidt keep pointing out.[4] AI's neural networks, machine learning, and data analytics make up just one piece of the puzzle, although a very important one. But there are also quantum computing and augmented reality, the Internet of Things (IoT) and Industry 4.0, Web3 and the metaworlds, genomics and blockchain, satellite constellations, nuclear fusion, and other space, bio, and energy technologies.

Their combined impact is so profound that I call their effects "*tech*tonic shifts." Like moving tectonic plates in geology, technological forces often are imperceptible. When the stress passes a critical threshold, though, they release massive energy and disrupt the world with often cataclysmic consequences.

Today, and continuing into the future, the new technologies that I just mentioned are rearranging most of our lives. Many of them are even growing exponentially. They are shifting the world under our feet, with a transformative impact on liberal and authoritarian societies alike. This then shapes how we master humanity's challenges, generate wealth and prosperity, and fight wars or maintain the peace.

My main background and expertise is in such emerging technologies. I analyze how they work, their impact, trends, and trajectories. Then, I assess how they affect the future of our societies, businesses, and lives.

But I am also a lifelong student and practitioner of international relations. Almost everything I do involves crossing borders of many kinds. Geography, nations, cultures, political systems, platforms, metaworlds, online spaces - you name it. To bring these two fields together, technology and international relations, and hopefully to contribute at least in a small way to preventing a large-scale war of choice, I wrote this book.

My main point is that techtonic shifts and innovations continue to transform our world and we have no option but to adjust. They open vast new opportunities for the world - but only if we do not repeat the suicidal decision of an early 20th-century Europe that had encountered a similar situation. Then, Europeans decided to engage in war among great powers. It eventually killed over 70 million people, destroyed large territories on three continents, and wiped away the powers and institutions of those that started it all.

In the mid-21st century, such a war between great powers would have similar results. Like then, its ferocity would surprise most or all of us because it would come at the heels of multiple decades of peace among the most powerful nations. And similar to the one of a 100 years earlier, a 21st century war would also be conducted in ways that we can hardly imagine.

It would be uncontrollable and unpredictable, with most likely far more horrible consequences than most of us think. War between China and the U.S. would directly affect everybody, probably hurting and killing countless lives and the near-limitless prospects of a generation. It would achieve all this primarily in the rich and seemingly invincible world of the war-fighting powers' homelands.

But too many who frame today's major conflict as one between China and the U.S. and their respective systems ignore that both are in some important ways more similar than they seem. The more important struggle is not *between* them but *inside* their societies. We in the West cannot defend a "free and rules-based liberal world" if we focus on the past while moving ever closer to mimic China's top-down and totalitarian methods. Rather, we must look at the future and use emerging technologies to re-establish a genuinely free and liberal rules-based system that empowers people. Skipping the terms "free" and "liberal" does not get us there. Rules alone are not enough.

Before we engage in ambitious strategies to militarily confront another world power like China, which seems most probable over political control of the island society of Taiwan, we should consider Mahbubani's and del Pero's references to the long arc of history. And we also must learn critical lessons from past world wars.

World War I teaches us that a great power war must remain an absolute last resort and not an instrument of policy. This is even more important to keep in mind when war escalates via automated reflexes or mechanisms.

World War II teaches us that modern war fighting tools can easily bring the fight back to civilians and into our homelands, wreaking havoc in unexpected and unprecedented ways. The personal and physical dangers for everybody's daily lives are huge. Too few in societies that have not experienced war inside their borders pause long enough to properly contemplate how modern technologies and asymmetric warfare can turn against them inside their home countries, or even inside their own homes. These people include almost all those living in China and the United States.

Both world wars also show that ignoring underlying trends and fighting systemic change is futile. Doing so can destroy wealth, and it usually does. At a high cost, it only temporarily holds up processes that already are in motion. This delay simply makes disruptions reappear more explosively. It also has massive costs of missed opportunities to improve our lives. Focusing on fighting and destroying instead of adapting and innovating prevents us from taking advantage of the abundant opportunities of technology-driven change.

To address all of this, I structured the book into three parts.

In **Part I**, I analyze the main geopolitical trends and their current and future trajectories until at least the mid-21st century. This sets the stage for putting war over Taiwan and its underlying challenges into the proper context that all too often is missing from today's foreign policy debates.

Digitization and space change the meaning and effects of *geography's* distances and features and makes them less relevant. It transforms concepts of Island Chains, Freedom of Navigation, and Chinese ambitions of becoming a hegemon over East Asia, Eurasia, or even the whole Earth. The world looks different when looking out from China. In Eurasia, its power is solidly balanced by Japan, Indonesia, India, Russia, the EU, and the United States. Its land borders are secure. However, the most populous country, largest economy, biggest manufacturing power, and dominating trading nation of the world feels threatened by the strongest navy of the world patrolling its shores. Neither its ships nor its own trade can reach the blue oceans without passing islands and traversing waters controlled by antagonistic forces.

But when concentrating on islands and the seas, both sides focus on the past. Geography and political borders matter little in space and in the digital worlds. Trade keeps shifting from physical supply chains to digital ones. And surveillance from space or via drones can limit any navy's actions, everywhere on the planet. All this upends geopolitics and repudiates traditional arguments and strategies.

D*emography* shows that the spread of population numbers between the U.S. and China will peak before 2030 and then shrink. The faster aging of a contracting Chinese population, while the U.S. is more likely to continue growing, will introduce new kinds of challenges. India's and, above all, Africa's population growth may soon create more geopolitical disruptions than China. This should affect U.S. priorities. America also has some important decisions to make about immigration. And prospective advances of medical and anti-aging technologies make all trends very difficult to predict beyond the next few decades.

About the *economy*, prudence suggests that we must consider China an equal with the definite potential to pull away and far ahead of the U.S. in absolute power. America's often encrusted bureaucracies seem increasingly inefficient when compared to China's technocratic and utilitarian innovations that frequently leverage capitalist forces. Those have delivered economic growth rates and an advancement of society that makes many people on the globe wonder whether Western ideals and systems can keep up.

In recent years, America's *soft power* has eroded while China's has grown. China built state-of-the-art world class infrastructure and pulled hundreds of millions out of poverty. In the United States, though, many around the world see globally advertised domestic hyper-partisanship, never-ending political campaigns with relentless personal attacks, and the brutal tearing down and demeaning of many of America's most successful and cherished institutions. They notice small but ever more radicalized self-anointed elites and often inefficient bureaucracies focusing on what seem secondary issues. All this reduces the U.S. 'global appeal.

We must also consider China a near-equal in *technology*. This has a decisive impact on geopolitics. Particularly for digital technologies, it is critical to understand the revolutionary and transformative impact of *exponential* growth rates as compared to traditional concepts of *linear* and incremental change. And the way our societies use and organize technology will be decisive for our future. At its core is the challenge of protecting privacy and ownership rights over digital products and data as the "new oil" or "lifeblood" of the 21st century.

In **Part II**, I confront the misguided notion that war between great powers in the 21st century can be a serious strategic option. In five chapters, I dive into what a war between the U.S. and China could look like, and what role technology would play in it. This examination is necessary because there has not been a great power war on planet Earth for three generations.

It becomes apparent that we in the U.S. put too much emphasis on our presumed technological leadership. We have many times used it against far inferior enemies in wars that we still could not win decisively. Rather, and partly *because* of our technologies, the way we organized our society makes our own homeland highly vulnerable. And our slow-moving military procurement continues its focus on either digital surveillance or outdated military systems and technologies. It does not leverage the full innovative potential of our society.

Militarily, and according to other geopolitically relevant factors, the U.S. is a clear *second* in East and Southeast Asia. On the military battlefield of the Western Pacific, there is no realistic way of achieving anything definable as traditional "victory." Rather, a war between the U.S. and China would almost be sure to *escalate*. This is so because there is hardly any realistic chance for the U.S. military to "defeat" the Chinese People's Liberation Army close to China. The only way would be by bombing the Chinese mainland.

But it is near-inconceivable that China would then not strike back. Instead, and just like the U.S. would react in a comparable situation, China could and would retaliate with new kinds of horrible weapons. They would be non-traditional, asymmetric, and probably non-nuclear, and target the centralized networks of our core infrastructure. Our homeland would then be under attack. With relatively simple means, it would be possible for China to deny our citizens access to electricity and energy, transportation, communications, goods, money, and even food and water. It could bring down nearly all critical components of our society. This would lead to widespread suffering across the continent. And America could do much of the same to China.

Part III proposes specific solutions to prevent war and devastation. To get there, we must understand that the 20th century's "rules-based liberal world order" does not exist anymore. Exponentially growing digital technologies have completely transformed its underlying conditions. This makes it senseless to cling to an idealized but bygone past. Resisting the *techtonic* and geopolitical shifts outlined in Part I will only lead to more pain and greater dangers, including war.

Outdated institutions and strategies prevent us from properly addressing the challenges of the future. This means that, today, America no longer can unilaterally enforce, or force, order for the world - almost anywhere, and definitely not near China.

To maintain peace in Asia and the world and to achieve prosperity, we must redesign our military, refocus it on defense, and equip it with new kinds of weapons and strategies. We also must remold our society and diversify and decentralize its underlying infrastructure to make it more resilient. Most importantly, though, we must build a *modern version* of a 21st century style rules-based liberal global system that sets a clear counterpoint to China. Its focus must be on the future and the new commons of the digital and space that are replacing the seas. Above all, this means the next generation of the Internet: Web3.

All three steps require us to accept change, welcome the future, and incorporate technology-driven opportunities in a way that re-establishes broad support inside and across liberal societies. For this, we must embrace and lead the digital transformation and take advantage of its opportunities. Encryption, protection of digital ownership rights, and decentralization of data and systems must become important ingredients to ensure the continued existence of free societies.

We must deliver peace and prosperity using 21st century tools, not 19th or 20th century ones.

If we do so more effectively than alternative philosophical world visions, like China's, the system we build will be stronger and more powerful. Others will then voluntarily adopt it. We will neither fight war nor destroy our society and the conditions for prosperity.

But if we continue to mimic ever more and larger aspects of China's system, we will lose because this directly contradicts our most cherished principles. And above all, if we fight a war with another superpower, we risk destroying it all, including our wealth and our opportunities.

War or wealth, authoritarianism or freedom, control or empowerment - these are the choices. Technology can deliver any of these.

Building a successful liberal future means trusting the bottom-up wisdom of the crowds. It also requires embracing modern technologies to continue improving and transforming our lives. Both are embodied in democracy and market-oriented principles. We can modernize them in practical ways if we put our minds to it.

We can and must give peace a real chance, and life, liberty, and the pursuit of happiness for all people.

Bert Kastel
New Castle, Delaware
July 2022

INTRODUCTION

The Lure of Suicide Amid Techtonic Shifts

1
Man-Eater

The Man-Eater of Taiwan

Man Eater. This is what they called him because that is what he did, almost literally. 30,000 of them. I visited him in the late 1980s, when my grandfather asked me to be his chauffeur for a weekend road-trip.

Unlike him, I had a driver's license, which until then I had not yet needed a lot. Because this was my first unsupervised multi-day trip, it was a big deal that my mother let me use her car. But it was her father's wish and just for an extended weekend. It also was a relaxed trip and unforgettable fun.

Our family has German and French roots, and my grandfather was particularly fond of France. And so, we leisurely drove south across the French-German border and into the Alsace. Our southernmost stop was a mountain, or a hill for those like me who have lived in the American West: the Hartmannswillerkopf, or Vieil Armand.

When you stand on its top, it is a genuinely pretty and peaceful place. Amidst mature and healthy forests, you can breathe fresh mountain air in pristine nature. You have a beautiful view into the

Rhine valley of the Alsace and beyond at the German Black Forest, and westward over the Vosges mountain range toward the heart of France.

And you also can see thousands of graves. Row after row of neatly arranged white crosses. Commemorating the about 30,000 men who died here over 100 years ago.[5]

This hill was not always pristine. It was the site of a protracted trench warfare battle at the beginning of World War I.

During that "Great War," technological advances had introduced new kinds of weapons at a mass scale, with hellish consequences for those attacking. But attack they did, nevertheless, using outdated tactics against revolutionary new kinds of weapons. Waves of men, row after row of them, attacked enemy lines. They did it in many ways similar to how armies confronted each other during the Napoleonic Wars a hundred years earlier. And so, they ran across unprotected open fields at well-defended positions and into the machine gun fire that mowed them down like a blade cuts through grass.

Scenes like these I contemplated while standing at the Hartmannswillerkopf. Such actions define the terms "horror" and "senseless." Each attack required a deliberate decision to ignore overwhelming evidence pointing at its likely fruitlessness. But fight they did, for honor, for a king or his cousin, for theoretical principles of glory and abstract concepts of empire. At such moments, a hill seems worth 30,000 deaths, as if the existence of civilization or humanity would hinge on it.

It is a complete loss of perspective. And so, 30,000 of the best young men on both sides died fighting over some hill in what was just the opening salvo of a world-spanning struggle lasting 30 years. In its course, *over 2,000 times as many people* died across our planet, on other hills, plains, waters, and in cities. Some of this may have turned out as bad as it did because of *how* it all began, on that specific hill and the many others. It resulted in what some historians call the collective suicide of Europe.

Killing People and Prosperity

Tragically, these actions were taken when an unprecedented pace of technological progress had just started to create well-being and prosperity for more people than ever before in human history. A still small but growing middle class had emerged in North America and parts of Europe, ready to spread across the world. Industrialization triggered scientific progress and technological innovations. A first wave of globalization let humanity glimpse an extraordinarily promising future. Automobiles, vaccines, fertilizers, electricity, newfound vast amounts of energy and the rise of consumer products had begun to transform many societies and lift the masses out of poverty.

Discounting such opportunities and aggressively resisting geopolitical change (i.e., Britain), or insisting on it (i.e., Germany), the then pinnacle-societies of humanity's civilization decided not to focus on maximizing the benefits of this new world. Instead, they took decisions that destroyed most of it! Completely preventable all-out global war blew up lives, assets, and the promises of at least one generation. It delayed progress for humanity by decades, almost everywhere outside North America.

A century ago on the European continent, the then-superpowers destroyed each other's heartlands, following preventable deliberate decisions. Today, and particularly in Asia and America, too many seem to have forgotten this lesson.

Again, during times of accelerating change and dramatic technological and scientific advances, many of today's leaders and intellectuals contemplate comparable actions and argue in similar ways as those self-assured, world-ruling European aristocrats of a century ago. Like the rulers of that previous *European* "world order," they seem unable to shake off outdated institutions and strategies designed in and for a world that no longer exists. And therefore, they also risk making the same kinds of mistakes, with probably the same kinds of results, or even worse ones.

At the beginning of the 20th century, politicians, generals and their commanding officers asked thousands and eventually millions of their countrymen to line up and charge against well-protected defensive positions in a war where defense ruled supremely. Farmers, bakers, teachers, and factory workers in uniform, with rifles in their hands, were asked to storm over open fields, cross barbed wires, run over mines - and directly into enemy fire and machine-gun bullets.

The consequences of this folly could not have been more disastrous. Little over 100 years later, though, ever more people seem to seriously consider great power warfare again. And this time, it could bring something even worse.

A Fake Dilemma: The Taiwan Trap

Today, many words, actions, and preparations point at China asserting itself ever more aggressively and challenging the order that America and the Western world had built after 1945. The most notable sign is China's pursuit of ruling Taiwan, taking it peacefully or by force. Politicians, academia, the media, and numerous experts all over the world keep expecting exactly this. For many, it seems inevitable.

The key question seems to be whether the U.S. military should fight and defend Taiwan or not. In the United States, most analysts describe a dilemma: if America refuses to fight, it will have left a democratic society being gobbled up by a totalitarian regime. If the U.S. fights, though, it will have been sucked into an uncertain conflict to defend a land that is not even a formal ally, against another superpower. Both choices seem to be equally bad.

But they are not. This is not even a genuine dilemma. Directly engaging in such a war would, on both sides, be as senseless as the one 100 years ago. Such a conflict would draw us into a war that quickly could engulf the entire region, or even the world, including each other's homelands. The option to fight is worse than the alternative, much worse.

At the same time, it would ignore that many of the conditions, parameters, and logic of what we would fight over no longer apply and make sense. Substantive changes are transforming the world under our feet, driven by emerging and exponential technologies. They are transcending old concepts of geopolitics and redefining how our societies function. We must acknowledge these changes rather than fighting them like those monarchs of the 1910s.

Even so, many among today's superpowers' academic and political elites are again suggesting similar actions as the European leaders of the early 20th century. On both sides (!), many of their arguments are about geography and the ability of their naval forces and commercial ships to traverse oceans without restraint. Others are about history or principles of individual freedom, political process, or societal systems. All are fair and logical enough, at least in theory.

But the crux of the matter is that all these try to guide our actions by describing a world of the past that is increasingly becoming irrelevant to the future.

As a result, Chinese elites push to establish control over adjacent islands for mostly symbolic reasons. This ignores that in the mid-

21st century, dominating or "owning" the Ryukyus, Paracels, or even Taiwan, has barely any substantive geopolitical, economic, or even military impact. In the 1950s, any actual threat to Communist China from Taiwanese Nationalists evaporated. And after China's hyper growth had started in the 1980s, Taiwan soon turned into a close trade and investment partner that today is dwarfed by the might of the communist People's Republic of China (PRC). Even so, today, many Chinese seem to consider it necessary to subjugate Taiwan's people even at the risk of global war in what primarily seems a symbolic move.

On the other side, in the West, elites urge preparing the United States and its allies to counter such a Chinese military move at almost any cost and get ready to fight on the potential battlefields of the Western Pacific. They ignore the bigger picture of how real freedom and prosperity must be won in the 21st century, and definitely how modern great power war will be conducted. And so, they urge their militaries to prepare for charging over open waters toward hardened battlefield positions with new and "game-changing" weaponry.

Although either side may win battles, China is the heavy favorite to defeat the U.S. forces in the theater. But as I will explain below, this also makes a dramatic escalation with extreme consequences likely, or at least possible.

Formally, such fights would be about planting flags or "freely" traversing oceans as if we lived in the 19th century, or about a small society's right to govern itself and make its own laws. But in reality, it would treat a small political entity, and geographical features like islands, island chains, and oceans, as crucial for future generation's peace, prosperity, and freedom.

This is nonsensical. Our future will be determined by the digital world and space. These are the main realms deciding the success of our principles. Space and the digital define the real "battleground," if such a term is even appropriate. Ignoring this would mean that, as so often during our history, we humans would fight the last war. More precisely, it would be the one of the 1940s, with weapons similar to those of the 1940s, to defend a status quo similar to what the world looked like in the 1940s.

Old Arguments and The New Commons

In the meantime, and on both sides, many of today's debates and even arguments sound like those of a hundred years ago, particularly in one way. They describe a zero-sum state of the world: "you win - I lose." Arguments center on military and traditional geopolitical aspects and not on healthy competition to provide better solutions for one's own people and for the world.

Systems and policies of both sides are also converging in ways that often mirror each other. Among these are restrictions of freedoms, near-total surveillance, and loss of privacy and digital security. They keep enacting heavy-handed top-down economic policies. And on both sides, they subjugate corporate institutions to politically derived social priorities conceived by small groups.

While technologies are transforming our societies with warp-speed, we freeze. Our instinctive reaction is to retrench and fall back on past strategies and institutions. Some of these are over 75 years old, though, created in and for a different world.

Our debates mention important principles, such as in the Western case those of "freedom," "liberalism," or "capitalism." Even so, they fail to acknowledge that the major challenges to such ideas are not from *other* nation states. Instead, the biggest danger is the rigidity of our thinking and actions *within* our own nations.

Many of our most cherished social and economic institutions are rooted in outdated concepts of the late 19th or early 20th century. We enshrined them in vast bureaucracies. Overwhelmed by technological disruptions, they adapt by throwing money at problems. This temporarily strengthens the obsolete, which then shuffles paperwork, engineers society, and pays for ever more administrative overhead.

But whether or not we are prepared for it: 21st century technologies are changing this world. They upend how people and nations create prosperity, organize societies, provide social goods, and even what the idea of a free and liberal world means in real life.

During discussions about war and peace, many people seem to knowingly ignore the trajcctories of crucial factors relevant to geopolitics. They only casually touch on such core aspects as demographics, economics, even geography itself and, above all, *technology*. Academics, media, and politicians then near completely disregard the impact of great power war on our civil society's digital and networked systems. This shrugs off the potential, or even likely, extreme price that humanity would pay almost immediately. With this, I mean the suffering and death of massive numbers of civilians in our homelands.

In a recent book, former U.S. Secretary of State Henry Kissinger, former Google CEO Eric Schmidt, and the Dean of the Schwarzman College of Computing at MIT Daniel Huttenlocher warn about the paradigm-shifting impact of artificial intelligence.[6] While they are correct, today's geopolitical and technological transformations go far beyond AI.

In the 2020s, the realms of the digital and of space enable the near-instantaneous global reach for anybody, anytime. They make it possible to create and transfer value or inflict harm from far away. Traditional political, geographical, and even legal borders and institutions can neither effectively limit space, nor zeros and ones, nor quantum states. And notably, all this then also alters most conventional concepts of military power and warfare.

Continued advances in digital and space technologies make defining and structuring *their* platforms crucial for our security, prosperity, and freedoms. The new world uses artificial intelligence, augmented reality (AR), the spatial Internet and Web3, the Internet of Things, 5G and 6G, vast satellite constellations and the automation of manufacturing to increase efficiencies and connect all people and their actions, instantaneously. Such new technologies require new kinds of answers.

Unfortunately, the political class in either society, but particularly in the U.S., does not seem to understand the degree to which the world has evolved and continues to change. The default behavior then is to continue as if things were as before. Like the Europeans thought in 1914 that they could fight modern versions of the Napoleonic or French-German wars, today's leaders look for guidance in 1945.

WWAD - What Would America Do?

Although I have just described and compared both the Chinese and the Western positions, I wrote most of this book from an American perspective and with an American and Western audience in mind. And my first question to them is basic. When our political and thought leaders demand that we must militarily "fight for Taiwan," what do they really mean?

It does not sound like most are talking about "mere" intelligence sharing, economic sanctions, military equipment, civilian supplies, or selective cyberattacks. Rather, they mean direct acts of war using our naval, air, space, and marine forces against those of China's People's Liberation Army (PLA). They advocate using U.S. weapons to shoot at Chinese weapons when Chinese weapons are shooting at Taiwanese weapons.

Considering the tactical, numerical, technological, and logistical context, such actions are highly unlikely to stay at sea or in the air. Rather, they will involve some variation of what has been called "integrated AirSea Battle."[7] This battle doctrine is supposed to blind Chinese digital support systems on land and on the water, target the People Republic of China's (PRC) operational logistics, and destroy PLA weapons in the air and on and below the sea. Such actions are to wear down Chinese offensive and anti-access/area denial (A2/AD) capabilities in the Western Pacific Region of Operations.

This then would, in theory, clear the way for large American military forces to cross the Pacific and join in the fight. However, a decade after the concept was introduced, the continuing numerical and technological shifts in China's favor have profoundly weakened AirSea Battle's chances of success. That is, barring its most extreme escalatory step: the U.S. must attack Chinese defensive positions on land *first*.

If the U.S. military does not do so, its military may inflict some damage, maybe even a lot. But it will mostly do the equivalent of running up the hill into machine gun fire without a realistic chance of success. So close to China, the odds are overwhelming stacked against American forces. To prevent a senseless bloodbath, the U.S. military must therefore consider breaking through fortified defenses via an attack at sites on the Chinese mainland. Doing so, however, will target positions near many of China's population and industrial centers. American bombs and fighter planes would hit the Chinese homeland.

Contemplate this situation in reverse and ask yourself, "WWAD." What Would America Do?

Yes, then all bets are off.

Brutal Honesty Before a Brutal War

Unless the above changes, claiming that we must fight as a matter of principle is equivalent to generals ordering an attack on fortified enemy positions in the trenches of the Somme or Verdun in 1917. Throwing away the lives of our best fighting men and women would reflect the same kind of cynicism. We cannot just *hope* that we may luck out when confronted with a massive imbalance favoring the other side.

We must also be upfront with our citizens before we ask them to carry burdens and pay the costs of such actions. Not explaining what such a war may entail for the American homeland would be inexcusable. It would be comparable to the effects of whipping up the masses by declaring "total war" in Berlin in February 1943. At that same moment, staggering numbers of American and British fleets of bombers had been gathering to target civilian life and infrastructure in their enemy's homeland. All may sound great in the moment's passion. It may even mask the possible or even likely grisly consequences. But it cannot evade them.

Simplistic positions of academics and idealists confuse theory and reality. They mix up short-term heroism with long-term success.

Obviously, nothing I describe herein is guaranteed to happen. Life is not deterministic. Humans make decisions in a complex world, setting in course a series of actions and counteractions. Nobody really *knows* what they end up with. As things go with human actions, including wars, specific predictions are extremely risky business and usually fail.

But the big picture, the visible trends, and the hard facts that I am describing in this book are real. They reveal what *can* happen and show what seems more *probable* than not. They also point at the possible or even likely consequences of our actions and what we must expect and prepare for.

Many decision makers and their influencers seem far removed from the real horrors of war, or at least they think they are. And somewhat ironically, the dominant positions of both major camps, China and the U.S., keep ignoring the same overarching truth: the limits of their power.

On the American side, too few seem to have internalized what it *really* means to move from a unipolar world back to a bipolar or even multipolar one. In China, too many seem to be drunk with the exaggerated implications of the recent rapid rise of their country to become one of two global superpowers, at least economically.

Almost all in a position of power today have grown up in a world where they never had to experience and face the most serious consequences of their decisions: destruction, deprivation, and the death of their friends and families. *Both* China and the United States have not fought a great power war in the lifetimes of almost all their people. For at least three generations, and apart from either minor or isolated instances, neither power's homeland has been battered by war machines, bombs, or other means of destruction. War is something they and their masses read about in books or watch on TV, not experience in person.

The human brain extrapolates experiences, the past. If all has been good during one's lifetime, it is too easy to take it for granted that this is exactly how life will continue. Consequently, assuming that oneself or one's own are not being put in harm's way makes it easier to send others to fight for what sounds high and mighty today, like nationalism and democracy. Tomorrow, the same may be called irresponsible foolishness when it ended up destroying proud nations and free societies.

The consequences of pursuing a principled stance detached from reality, particularly without clearly communicating upfront what it would entail, resemble the beginning of World War I. This is about the *decision* to engage in active warfare.

Once the fighting starts, though, it can quickly escalate. It would then more closely look like World War II on the European and Asian battlefields, with its direct and horrible consequences for military *and* civilian lives in the combatant's homelands.

Not bombs and tanks, but 21st century digital technologies will make sure of it.

From Ambiguity to Apocalypse

We must, therefore, confront the life-changing aspects of our decisions and positions now. When the shooting starts and the machinations of war take on their own life in a matter of hours and days, it is too late to start thinking about the big questions. Media cycles, algorithms, tribalism, and emotions of the moment do not favor a calm weighing of strategies, what-ifs, and the long-term consequences of our actions.

Once war begins, we will no longer have time for detailed and deep deliberations and debates that analyze for our citizens what such combat in Asia really can mean for *them*. The difficult-to-calculate overall impact of continued change through new technologies adds even more volatility. At home and on the battlefield, technological revolutions compound and transform our lives. Everything becomes more digital and more integrated. In war, this means that lethal actions literally can happen at the speed of light. This volatility and velocity will further complicate and therefore de-stabilize each side's calculus. They will also scale up the deadly consequences when *systems* get attacked.

From the U.S. perspective, here is the reality:

> A U.S. war with China is hardly going to stay limited. It will not be fought far away against technologically, economically, and militarily inferior enemies like those America has encountered since World War II. Such a war also is unlikely to be conducted like the Cold War, and definitely not under the same conditions. With no buffers between the probable military theater and the Chinese mainland, far-reaching decisions may have to be made within days, hours, maybe even minutes. Just like during the run-up to World War I, institutionalized automatisms and tactical actions on the ground are going to generate their own dynamics. Worse even, many decisions may already be automated, and soon they may even be made by AI algorithms, near-instantaneously. And all the while, *the ability of the war-fighting powers to cause serious harm to each other's homeland with conventional or asymmetric non-nuclear weapons is higher than ever before.*

We give too much credit to strategies that worked in different times and situations when a strong, even dramatic, military imbalance existed between a U.S. superpower and almost all its enemies, including an underdeveloped and then clearly inferior China. Our strategies and most war fighting tools still resemble those used in wars three generations ago, under completely different conditions.

The publics in China and the U.S. seem to be unaware of the explosiveness of these facts. In either case, their citizens assume obvious military superiority while being oblivious to the possible, or even probable, real cost in blood and treasure.

This is understandable because a genuine great power war has not been part of our lives. And therefore, most people have a somewhat distorted picture of war.

Correcting Warped Images of "War"

Judging by what we see in the news and movies, war usually invokes images of insurgents or terrorists fighting uniformed troops in Middle Eastern of African countries. At other times, it is precision explosives being launched by fighter planes or dropped by bombers. And sometimes they are World War II-like battles fought between tanks or among aircraft carriers. As different and pervasive as these depictions are, they have two commonalities.

They all take place far abroad, and they all are wrong.

Granted, these things have happened in many parts of the world and still do. But in the 2020s they do not reflect some of the most impactful consequences of war for Americans. They do not describe how modern large-scale war between great powers is likely going to be conducted. Most importantly, they miss any indication of real hardship and destruction inside America (or China). We hardly ever hear specifics about widespread devastation and suffering in our homeland, or of large numbers of civilian casualties.

Because these pictures of past, limited, and far-away wars are so pervasive in our minds, though, they are the main backdrop for opinions about foreign policy and decisions on military strategies and actions. These are girded by a sense of superiority of having the most effective and powerful military in human history. We hear of awe-inspiring weapons without match, and a near-limitless defense budget that suggests an aura of invincibility.

Such perceptions have real consequences. Spawned by Beltway think tanks and academic experts, advocates make comfortable and heroic assessments and recommendations. Decision makers reach profoundly consequential decisions based on exalted ideals, lofty aspirations, and a simplistic selection and interpretation of facts.

And so they downplay or ignore uncomfortable hard truths, selectively look at history, and override common sense with abstract theories and wishful thinking. Take your pick: Vietnam, Iraq, Afghanistan, even Chile, El Salvador, Somalia, Libya, and Iran. They all had unintended consequences and negative results. Today it is the South China Sea. And above all: Taiwan.

Often, and this is where many of the above pictures come from, we then end up fighting wars in small countries far away. We do it for proud and sometimes complicated reasons. Often these are to spread or defend freedom, protect a world order we still call rules-based and liberal, conduct forward defense, prevent dominos from falling, or to build nations.

At times, we fight simply because we are convinced to be the successful and civilized good guys and the others are not. We tell ourselves that we spread and protect universal truths. And this may even be correct. But then we also say that imitating our specific system of society is so important to others that they want to have the same way of life, at any cost to *them*. At that point, we become convinced that we therefore must support them in their endeavor, no matter what the cost to *us*. The others, the dark side, the enemy, we comfortably assume, must always have more selfish and nefarious motives.

And so, despite our past experiences, Washington keeps following the same playbook over and over again. It whips up and magnifies the threat from adversaries. Then it tells us that already the first battle would determine the fate of civilization. Instead of a fight over a hill, it already seems like the final scene of The Avengers. To cooler heads and those looking for alternatives, they mention "Munich" and cry "appeasement" while touting the success of the Cold War. And this simple recipe seems to be sufficient to justify pretty much any military or foreign policy action. So let's confront China (or Russia) now rather than later, and beat them back while we can.

But I see overwhelming evidence to read the facts differently: world war with China would not likely be the way it is being sold. It would not stay far abroad and limited. If it happens, it will come to a town near you, and to your own. And it will be a war like Americans and most of the world have never experienced before. It will not be a World War III as a continuation of wars that ended 75 years ago. We will not simply be adding some high-tech wizardry to old-school weapons on far-away battlefields.

It is not guaranteed by any stretch that war with China would not go nuclear and physically destroy our cities. But even if we assume that nukes will not enter the fray, the connectedness of our modern world can easily turn it into the most monumental conflict of history. Such a war would be truly outsized compared to any other experienced by almost all people alive today. It would quickly affect billions of lives, almost no matter where people live. Merely thinking through the implications of collapsing just-in-time supply chains for critical products and goods *inside our homeland* suggests that this would happen. And this could be triggered by blowing up, with relatively simple means, a limited number of pipelines and a few dozen substations on our national electricity grid. Consider empty stores and having to live without electricity, gas, Internet, and food.

Imagine it as a World War 3.0, because modern technology is a big reason why such a war's impact would be so dramatic.

Imagining and Developing Solutions That Protect Us in a Transformed World

In a lot of today's public debates, we hear about a need to incorporate empirical data, of following science with an open mind, and not putting beliefs ahead of the facts. Foreign policy seems to be a big exception. America often seems able to bend reality to aspirations. And so, international affairs have become a tempting field of big dreams for many idealists, academics, and politicians. They often argue as if the presumed superiority of an ideology, of the idea of "life, liberty, and the pursuit of happiness," would be enough when reality does not add up the way they want. Too easily, well-meaning actions then end up destroying lives, limiting liberty, and delivering agony.

This is likely rooted in the vast power difference between the United States and every other nation. Simply put, we in America can afford to pour trillions of dollars down the drain while fighting in foreign countries. We can pay for the largest and most expensive high-tech weapons, like nuclear-powered Ford-class aircraft carriers, SSN(X) attack submarines, or next generation stealth fighter planes and bombers.

Our budgets and wealth afford us an unmatched global reach in cyberspace, too. We achieve this through NSA surveillance, Crypto AG tools, 5-eyes, 9-eyes, 14-eyes intelligence sharing agreements. Highly effective means also include tax and anti-terror legislation-driven financial reporting requirements that can reach nearly anybody, nearly anywhere on Earth. And we round up our capabilities through "public-private partnerships" with surveillance and data behemoths like Google, Amazon, and Microsoft.[8]

The monitoring of global communications and collection of financial and commercial information merges with our high-tech weapons, global alliance systems, and international bases. They help us keep wars far off our shores. It also limits the actual numbers of those wounded and killed in action, with annual totals rarely exceeding the hundreds. In Iraq and Afghanistan, even the peak of about 1,000 deaths a year in 2004 and 2007 were just a fraction of those who died in car wrecks or through falls.[9]

The military personnel and their families pay a price. But in our daily lives, war hardly affects any civilian American seriously. We barely even notice.

We fund all this with sheer unlimited budgets. A trillion dollars does not create any kind of headache for the U.S. We don't even need to "find" the money somewhere, we really don't. After all, we *own* the US dollar, the world's reserve currency. We not only create our own money through simple bookings by our federal bank. Unlike with any other country's currency, nearly everybody uses and wants our money, too. And so, with defense, we all subscribe to modern monetary policy: we can and will print as much as we need or want. Deficits don't matter.[10] Money never is the issue. Or at least it never has been.

Consequently, since at least Woodrow Wilson, the foreign policy leaders of the U.S. have used America's unmatched economic wealth to make aspirations and "normative" statements, and therefore *ideology*, a pillar and often enough centerpiece of foreign policy. They became the foundation for international relations strategies, policies, and actions, including those resulting in war.[11] Wealth and hard power afforded us the luxury of interpreting our interests broadly.

But now this does not work well anymore. We no longer are the undisputed number one economically and, in some theaters, also militarily. Rather, we must be much more cognizant of our limits than at any time since we entered the world stage over 120 years ago during the Spanish-American War.

Adjusting and taking advantage of technological shifts would make us economically stronger and let us continue to lead. On the other side, fighting systemic change, as we seem to do right now, always is a losing proposition in the long-term. We can keep wasting resources on it and hold on for a while because we are rich, like those European monarchs a century ago. But eventually, even we will run out of resources. Change will come. And therefore, we no longer can afford to waste precious resources, lives, and goodwill by clambering to the past.

Doing so would weaken us and eventually risk the fulfillment of our idealistic aspirations. Like most other people in the West, I don't like that there may be limitations for our ability to put principles first. Without power, though, the institutions and strategies we use to promote our principles no longer can shape the daily lives of others. And therefore, I'd rather make hard and uncomfortable decisions that consider our current limitations and make us more powerful in the long term.

Reality must come first to enable achieving our ideals. Reality first, then ideals. Sequence matters.

And technology.

2

Past and Reality

Technological Change and its Impact on Warfare and Geopolitics

Our planet has not seen a war among great powers for over 70 years.[12] The big nations have not gone full blown after each other for a very long time. Therefore, and luckily, most talk about such warfare remains theoretical. Almost none of us know the horrors of large-scale global war out of personal experience.

But there are things we *can* know about what a future world war could entail. Above all, warfare will hardly be limited to the kind of weapons that were used during World War II, and in subsequent smaller conflicts. New and powerful *types* of weapons have emerged that do not fit into traditional categories. And the world has been there before.

World War I was not decided by old-style weapons and strategies of the 1810s and even 1850s that marched, in formation, tens of thousands of soldiers into battle.[13] Rather, it ended through the slow bleeding to death of an entire generation, millions of people, in trench warfare dominated by machine guns and chemical weapons.

Soon after, in a continuation of what had turned out to be a 30-year struggle, neither of these decided World War II. Instead, it was fighting machines that had not existed in a meaningful way just 25 years earlier: aircraft carriers, tanks, fighter planes, bombers and, ultimately, nuclear weapons.

Today, these same 75-year-old weapon systems still dominate our armed forces and our image of war.[14] And sure, during an armed conflict between the U.S. and China, they would be used. But they are unlikely to decide a war among great powers in the 2020s or 2030s.

Merely 25 years ago, the *Internet* hardly played a big role in any person's life. In the meantime, technology has transformed all corners of our societies, work, and daily lives. The impact of mobility, digitization, and new forms of networks has accelerated change and profoundly altered the shape of our world and of all societies on our planet. This will continue during the next decades, probably even speed up. With this backdrop, traditional thinking is not enough when the full impact of artificial intelligence hits humanity, in combination with augmented reality, biotechnologies, digitized supply chains, ultra-fast 6G communication, and quantum computing, powered by a trillion sensors, mesh networks, and vast satellite constellations.

Whether or not we are ready, though, all this profoundly changes how international relations and geopolitics work, how warfare is conducted, and even what the words "war" and "peace" mean.

Technological advances do not simply give weapons new capabilities. For example, they don't just make them more precise, stealthy, or automated. Such frightening changes certainly continue to happen. But if we were to stop at this point when contemplating the subject, we would miss the larger truth.

Completely new types of weapons and warfare can inflict far more severe and widespread damage, suffering, and casualties than many of those we see in our militaries' arsenals. Going far beyond common cyberattacks, these unconventional weapons exploit the digitized and networked way our societies operate in the 2020s. They can target our infrastructure and systems, deny food and energy to our people, and fry our computers. And they can stop our cars, planes and trains, trigger pandemics, and create suffering on a scale we rarely have seen before.

The decisive battleground of the next great power war will be in and about this digital world. It will be about the infrastructure and platforms that define and enable our society's functioning. They have become the new *spatial commons* of the digital realm and of space. Not the oceans, as in the past, must be open and secure to ensure the continued existence of a genuinely free world. Rather, it must be above all these new pathways for communication, computing, and commerce.

The New Spatial Commons: Digital and Space

This has very specific consequences for international relations and how we approach war and peace. Instead of firing old-style weapons at other militaries and drudging through regional old-style fights nobody can win, we must focus on what really counts in the long term. Above all, this means being able to defend our society and strengthen the actual infrastructure and capabilities that the *future* of our world depends on.

We must therefore design and shape the platforms of the digital world and space to reflect the core principles that underpin democratic and market-oriented societies: decentralization and bottom-up empowerment. This can build on recent technological and economic advances, for example distributed databases and self-organizing mechanisms of the Web3 (or "Web 3.0"), and autonomous systems. In principle, their empowering of individuals is intuitive for us in the West. We already reflect decentralization, self-organization, and bottom-up empowerment in our society. In politics, we do it through democracy, and in economics through self-organizing markets.

Obviously, though, we never got markets and politics working perfectly. Loosely borrowing from Churchill, democracy and capitalism are often dysfunctional and possibly the worst forms of organizing societies, apart from all others that have been tried.[15] Therefore, it is worthwhile to improve on their often old and imperfect implementations without discarding their principles altogether. But we must adjust. As technologies keep transforming our society, most of our institutions will also have to adapt anyway, maybe even be replaced. They must differ from the old ones because they require new tools to address unprecedented sets of challenges and opportunities.

Examples of old institutions are the structures and processes we use to generate and distribute energy and food, and how we manufacture and deliver goods. But we also must change the way we organize our data flows, use financial systems, educate ourselves, secure our society, and conduct our foreign relations. Even the role of nation states themselves, and what "inter-national" relations mean, must consider how most cross-border transactions are nowadays carried out in practice. Usually, they bypass any direct involvement of governments and ignore many kinds of borders. People communicate digitally, establish virtual social groups, and conduct commerce with little thought of nationality or geographical location.

This need to reinvent institutions also applies directly to matters of military and national security. For example, even improved versions of old-style 20th-century weapons often are inferior to 21st century ones. Old-style weapons are easier to attack or defend against, because they are larger, slower, and less flexible. Smaller, distributed, decentralized, and autonomous weapons systems create asymmetric advantages against them. And for the same reasons, many of the old and way-too-centralized ways of organizing our society are vulnerable to infrastructure attacks. Among these are the ones providing energy, communications, or distributing food and goods.

Over 200 years after the industrial revolution began, though, it continues to hold us back from adapting in the most effective way. We still instinctively seek efficiencies through hierarchies and aggregating power in the hands of the few. Although we pay lip service to individualism, all too often, we fail to turn our principles consistently into practice. As a result, in markets and in politics, we frequently don't trust the bottom-up power of the people as much as the top-down rules of "elites" in business, politics, and science.

The rivalry between the United States and China reflects this challenge in multiple ways. It is not just about better ideals or philosophies, but also and above all, *about better institutions, better systems, and better outcomes for people.* We must continually prove to ourselves and to the world that our answers are better, and that they deliver better results than centralized, elitist, and autocratic ones.

These are not abstract statements. China's growth had tremendous benefits for its population. Like America's advances before, this also has enhanced the global appeal of the Chinese political and economic model and of its society because it indirectly also increased the well-being of many other parts of the world. Growth in Brazil, South Africa, and Australia, for example, relied on Chinese demand for raw materials. And businesses and consumers in most other countries benefited from the lowering of manufacturing costs and the emergence of a new and massive Chinese market.

Decades of economic success of China's version of development and its recently ever more radical turn toward techno-totalitarianism blur the picture, though. By looking at power centralized in the hands of the leaders of the Chinese Communist Party (CCP), it can seem like this political system delivered the economic results. It may then appear that top-down planning and totalitarian utopias would achieve superior results. [16] But this ignores what happened in China one level below. There, it was the unleashing of market and capitalist forces that drove the advancement of many parts of the Chinese society.[17]

To advance rapidly, China may not have applied political competition (democracy) but it definitely used economic competition (capitalism). Misunderstanding what really triggered Chinese economic expansion, and trying to copy the wrong parts, can then be risky or even dangerous.

This misinterpretation threatens everybody's individual freedom. Systems matter. We cannot have it both ways: top-down directives and centralized control on the one side, and individual freedom and bottom-up empowerment on the other. Therefore, America and the West can and must differentiate themselves more clearly and prove they can deliver superior results. Opportunities generated by new technologies create a chance to do so and confront threats for the future without imitating the ever more totalitarian measures that accompany recent Chinese advances.

Freedom, Liberty, and Empowerment

The all-important underlying struggle toward individual freedom and empowerment is far from over. But that struggle takes place not only, maybe not even primarily, *among* nations but also *inside* them. Varying degrees of limits on freedom exist in nearly all countries, as intrusive surveillance by governments and private organizations, centralized police powers, the elimination of privacy, eastern-style and western-style social credit systems, centralization of communication, censorship and the deliberate manipulation of people based on massive amounts of data collected about their personal lives and habits.

Such perils even exist inside our Western societies. Three individuals control the by far most massive databases ever established, documenting nearly all aspects of the lives of billions of people. These three, Larry Page and Sergey Brin of Google and Mark Zuckerberg of Facebook, control majority voting stakes in companies that generate almost all their profits from collecting data to build personalized profiles of every person alive.[18] *Each* of these profiles contains millions and billions of data points, documenting near everything about each one of us. These companies also work closely with governments, including the American one. That is the same American government hell-bent on "connecting the dots" via the Patriot Act, NSA snooping, and meta-data collection. It also is the one being reluctant to introduce privacy protection legislation preventing private organizations from doing what the government itself could not get away with.[19] As a consequence of the latter, Western governments don't spy (much) on us. But their business partners do, with impunity.

Human organizations, including nations, never are perfect. Therefore, and particularly considering today's upheavals, we cannot afford to stand still. Winning the rivalry with the Chinese regime requires continuous improvements of our own society. It is not about winning a traditional war but about delivering better results and more clearly differentiating our political and economic system instead of imitating the one we supposedly oppose.

For that reason, America and the West must also be extremely reluctant to fight with military weapons. It is easy to talk tough, but we must not get carried away in emotional moments. We must question much harder *whether* and *when* we have to stand up to fight with killing machines against aggression. We must remain aware that global systemic struggles, above all, are intellectual and innovative competitions. They will likely span decades and far exceed the impact of the first battle and of the opening salvos.

This is how it was in World War I, World War II, and during the Cold War. And throughout many of history's fights for hegemony.

How Technology Plays into This

Many note that militaries tend to fight the last war. Few realize, though, when their own thinking is similarly rigid and their imagination too limited. Likewise, stuck in visions of past glory and recent success, and misreading much more decisive underlying trends, both superpowers view the world using outdated geopolitical concepts. They look at the past and extrapolate it into the future. Meanwhile, both seem convinced of their own superiority.

Our current generation's privilege of decades without bloodshed among great powers left the large conflicts of the early 20th century as the only visual images of such a war. Not unreasonably, those envisioning a major conflict look at this past for guidance. But unfortunately, they then also mostly keep funding the weapons of that past, the aircraft carriers, fighter planes, submarines, and bombers to protect themselves and hit other nation's homelands. They build missiles and defenses against them and prepare for some form of nuclear exchange.

Recently, ever more of our leaders and thinkers start contemplating cyberattacks. Most only scratch the surface, though, and casually insinuate that these would be little more than a series of ransomware attacks and compromised databases.

We measure economic might based on how we achieved *past* Gross Domestic Product (GDP) numbers and growth rates, and military power on such conventional and nuclear weapons that succeeded *in history*. This thinking creates geopolitical constraints using concepts that no longer apply to the same degree, or even hardly at all anymore. And so it assumes that commercial trade requires the physical shipment of products across oceans, and that we need traditional naval and air power to ensure this kind of commerce remains possible. We call this ensuring the freedom of navigation.

Specifically, for the flash point number one in East Asia, this then grossly exaggerates the role of Taiwan. The Taiwanese society only has a population size of about 1.7% of mainland China's.[20] Despite that, and because of old-style thinking, the People's Republic of China concludes that it must strike at Taiwan and the neighboring small islands to secure unmitigated access to the blue oceans. This is self-deceiving, because not even control over Taiwan could achieve that goal. And the U.S. concludes that it must prevent such actions at all costs, although the tangible effects on its security and prosperity are relatively minor either way.

The fundamental problem with these positions is not that both are wrong, although I am convinced they are. It is that they ignore the larger context and make simplifying assumptions.

Great power war in the mid-21st century will be fought with new weapons, not better old ones. It will use the networked digitization of nearly all aspects of our society. Tackling infrastructure can inflict disproportionate pain. Asymmetric tools can directly impact the most vulnerable bottlenecks and building stones of our modern society. These are the electricity grids, the pipelines, the water supply, the domestic supply chains of food and commodities.

We must move away from our focus on the old playbook of land, sea, and air power. Instead, we must protect and leverage the new spatial commons of the digital world and of space. The same accelerating technological change that keeps reshaping our societies also makes it irrational to stick to outdated tools and strategies. It makes no sense to think or act as if we were living in the mid-20th century.

This is most important for the current stage of still accelerating technological change. As Friedrich Hayek and many others across the political spectrum and across scientific disciplines have described, centralized mechanisms cannot properly consider the impact of billions of interactions. They cannot do so, neither for specific atoms, nor for the lives and priorities of people.[21] In our societies, no single person or sub-group of people can do that either.

This is a tough point to accept for many educated people and powerful "elites." But trusting the innovative power of our citizens is key to an even brighter future. To benefit all, only such trust can fairly and effectively adapt our infrastructure and institutions. A decentralized world, like the one technically envisioned and being built as the new Internet 3.0 (or "Web3"), can make society function better in the future. It empowers all people, shifts old paradigms, and alters everybody's calculus.

This should be easy for those living in a liberal world. We already believe that the system of a decentralized, open, and organically advancing society is the critical means to deliver on the needs and wants of humans and their communities. In general, *more* such liberalism is the answer, defining a world that is more rules-based and free, and less centralized. This is the opposite to aggregating power in mostly self-anointed political, scientific, or technocratic elites of any kind. These reject, by definition, the empowering of individual's pursuit of happiness and the subsequent unleashing of their innovative power.

But there is one crucial condition for any of this to work: We must avoid a new world war, a deadly large-scale great power conflict, a bloody third Big One.

To build a better world, we can and must focus on the abundant opportunities that we can realize together. A major war between great powers, and the destruction, death, and devastation it is likely to deliver, would achieve the exact opposite.

We cannot advance the world if we destroy it.

Competition, Rivalry, and War?

About a decade ago, living in Singapore, I regularly attended a speaker series organized by Singapore's Rajaratnam School of International Studies (RSIS). Already then, more than one doubling of China's GDP ago, participants seemed acutely aware of the dramatic shifts happening to Asia's power system, and the subsequent geopolitical implications. Few jokes could evoke as good a laugh among a highly educated audience as suggesting that the United States could stay the dominant power in Asia. The main arguments were the 10% difference in annual GDP growth rates, the apparent discrepancy between the U.S.' words and deeds, and America's increasing practical irrelevance for Asian's daily lives.

Or as many sports fans in the U.S. like saying, "the stats don't lie."

So what are these statistics, the facts that show that America's geopolitical position in East and Southeast Asia has turned it into that region's number two? In the next chapters, I will start answering this question in more detail. I will dive into the general geopolitical context of great power rivalry in the Western Pacific. For now, only a few general points.

The above jokes were made in Singapore, a friendly and aligned nation, quasi-allied even, and among people overwhelmingly sympathetic to the U.S. and an economic liberal international system. Attendants of RSIS talks are also well-informed about international affairs. Therefore, what in retrospect may seem like prescient prophecy was little more than a simple and even conservative extrapolation of long-term trends.

Using basic mathematics, it was then already clear that China was not only catching up but rapidly surpassing, even eclipsing, the U.S. as the dominant power in East and Southeast Asia. A decade ago, most participants seemed to interpret the economic numbers and their personal experience in the region even more starkly. China would not need to surpass or eclipse the U.S. because it then already *was* the Number One force triggering change in Asians' lives.

Even in the 2020s, though, this message still has not sufficiently permeated the foreign policy circles on the other side of the Pacific, within a United States that, overall, probably still is the world's most powerful country. In the short time span between 2010 and 2020, though, China's nominal GDP has increased from roughly 40% of the U.S.' to approximately 70%.[22] Far more than any other power ever had managed on the Eurasian super-continent, China has grown to become Asia's, or at least East and Southeast Asia's, near-unchallenged regional economic hegemon. According to the World Bank, China's share of GDP reached in 2020 nearly 50% of Asia's total, and about 58% for East and Southeast Asia.[23]

In many respects, though, this goes beyond the economy. Despite bold claims of the political caste in the U.S., it appears that China has pulled even in some fields relevant to advanced and emerging technologies. Critical for a potential war, the same applies to China's military capabilities, particularly near China's coasts. This is for a country with the largest population on the planet, the largest manufacturing power, the most extensive infrastructure and an average economic growth rate that until 2020 consistently exceeded America's by 5% to 15% for about four decades in a row.

Many of those advocating for "defending Taiwan," "denial," "deterrence," and "containment" seem to understand. Then, though, and inexplicable to me, they usually discount or ignore the above realities. In politics, this is called "spin." Their selective views and interpretation of the facts then leaves them with lots of room to promote risky strategies and actions. But such a disconnect puts us on cruise control toward a major military conflict. Simply ignoring facts on the ground, empirical data, consensus interpretation of tactical capabilities, economic trajectories, and military trends can neither prevent wars nor win them for us.

First, we must face reality. So let's look deeper, and at more facts. A handful of geopolitical factors provide the background for everything we talk about. Here, "geopolitics" is not a loose synonym for international politics, as, for example, defined by the Encyclopedia Britannica.[24] I will use it more narrowly, closer to what Merriam-Webster calls the "study of the influence of such factors as geography, economics, and demography on the politics and especially the foreign policy of a state."[25]

Our goal here is to lay the foundation of understanding the profound shifts of geopolitical conditions impacting the perceptions and relations between the United States and China. This will help us answering a question posed by Australia's former Prime Minister Kevin Rudd in his 2022 book "The Avoidable War."[26] He calls it "the single hardest question of international relations of our century: how to preserve the peace and prosperity we have secured over the last three-quarters of a century while recognizing the reality of changing power relativities between Washington and Beijing."

A century is a long time. But if we answer this question in the wrong way today, for many it could be over sooner than anybody likes.

It starts with geography.

Part I

GRAND ILLUSIONS

Techtonic Shifts and Geopolitics in the Middle of the 21st Century

3

Geography

The Past and the Future

How long would it take an American fighter jet to fly from the continental United States to Taiwan? And how does this compare to a Chinese fighter plane crossing the Taiwan Strait?

A few facts help answering these questions. On the surface of it, the calculation seems easy. The distance between Taiwan and Seattle is slightly above 6,000 miles (ca. 9,650 km).

Assuming the speed of a commercial airliner, it would take an American fighter jet from the U.S. West Coast about 10 hours to reach Taiwan. If the jet could fly at Mach 2, it would take about 5 hours.

Achieving the equivalent would be much easier for a Chinese fighter. Taiwan is only about 100 miles (ca. 160 km) from the Chinese mainland. This is 1/60s of the distance to the U.S. mainland. It means that however many *hours* it would take for a U.S. fighter plane, that many *minutes* it would take for the Chinese equivalent, or for a supersonic missile.

5 minutes.

The above is theory, of course.

In the real world, a fighter jet has an effective operating radius of just several hundred miles. To cross the Pacific Ocean, it would need to be refueled multiple times, slowing down each time. Even more likely is that an aircraft carrier would transport it. The entire trip would then take anywhere between 12 and 30 days.

But if there were a war with China in 2022, chances are that it would never arrive.

As we all know, tectonic plates move slowly. For the eyes of humans, therefore, the general shape of Earth's continents, islands, mountains, oceans, rivers and lakes hardly ever changes. For millennia of human history, they have been the most immovable enablers and limitations for humanity's activities and the power of their societies.

Until recently.

20 years ago, a motley crew of Middle Eastern terrorists creatively overcame geographical limitations and political borders to strike targets over 6,000 miles away, in America. Their actions triggered massive destruction and extensive warfare that eventually led to the death of hundreds of thousands, cost trillions of dollars, and directly and indirectly spanned most of the globe. This is child's play, though, compared to what powerful nation states can do to each other in the 2020s.

In the past decades, modern technologies and their digital underpinnings have connected the world and all people across vast distances and geographical features, blurring ancient boundaries. As a result, oceans don't protect as they did in the past. Enemies can penetrate borders and cause dramatic physical harm without weapons that would be recognizable to people of the mid-20th century. Some of these weapons are digital, others space-born, and many combine both. They have in common that they ignore physical distances on Earth's surface. Geographical features like mountain ranges, deserts, and oceans cannot limit them.

More general consequences of digitization create previously unheard-of capabilities of new weapons. Among these are AI-guided autonomous drones, and swarms of them, or bioweapons taking advantage of advanced "gain-of-function" research. Such capabilities have entered the picture in a way that is ever more difficult to assess reliably and is largely unencumbered by political and physical borders.

In some ways, geography still keeps affecting power relations and military options. It is easier to move armies over plains and navies across open oceans than to make people, ships, and tanks traverse or circumnavigate mountain ranges, jungles, or archipelagos. This simplifies some forms of warfare and imposes some restrictions in other ways.

Space and digital technologies have removed many of these limitations, though. They profoundly transformed our tools and strategies for offense and defense. And therefore, we must completely rethink the way we approach the question of global power rivalry in the light of modern warfare.

To specifically assess what this means, we must start with a more traditional look at geography. This helps us to better identify how recent technological changes have impacted the question of war, peace, and warfare between the United States and China.

The Lay of the Land: Oceans and Chains

Geography affects a country's security in many ways. Clearly it did, and still does, establish limits and present opportunities for moving militaries and their war making tools into battlefields so that they can fight other nation's forces and conquer territory. But for matters of national security, it goes far beyond military weapons.

Nature and its features also determine where and how a nation produces or gets its food and other products, and therefore how it can sustain its population. The availability and location of natural resources dictates the supply of energy and raw materials. Geographical features and shipping routes shape the logistics of domestically or internationally transporting energy, raw materials, and products.

Geography also affects specific military options and strategies, including the ability to directly "project power." This euphemistic term denotes the positioning of weapons close to a potential or actual enemy's borders and the threat to destroy valuable military or civilian assets on the other side of that border. It thus can make others do something they otherwise would not.

The U.S. achieves all the above through its continental scale and its unrestricted access to the oceans, supported by aircraft carriers and the world's largest navy by tonnage.

Things are very different for China.

A careful study of a map covering East Asia and Southeast Asia quickly reveals one of the major issues discussed in this book. On the one hand, most Chinese borders seem secure with plenty of "strategic depth." This term refers to the idea that other powers would have to cross large territories with conditions unfavorable to them before they could reach the core Chinese lands. China has such strategic depth mostly to its north, west, and south.

The situation in the east and southeast is different. This is where the oceans are. And even as the world's leading trading and shipping nation, with the largest navy by numbers,[27] China has no uninhibited access to the blue oceans. To reach open waters in its east and southeast, Chinese commercial and military fleets must pass through areas controlled by nations allied or closely aligned with its most important strategic rival, the United States.

Before we analyze this specific dilemma in more detail, though, we will stay on solid ground and on the mainland. China shares land borders with 14 countries. Most of these are far from the industrial centers in the northeastern and eastern heart of China, its bustling coastal cities and those along its primary river systems, and the manufacturing regions that stretch into the country's south. These industrial and population centers are separated from most of China's neighbors by rivers, mountains, and deserts in the north, west, and south.

Besides supplying raw materials and energy, the countries to the north are scarcely populated and do neither pose significant threats nor market opportunities. Mongolia only has a population of barely above three million people, and the Russian Eastern Federal District below eight million, living in an area the size of about 80% of all of China.

Further west and in the south, except toward Vietnam, deserts and mountains make logistics and commerce challenging because of difficult terrain and remoteness from essential markets.

As long as the vast regions of Tibet and Xinjiang remain under Chinese control, China is therefore strategically safe in its west. Deliberate policies aid this situation. It is difficult to know for certain, but a strong influx of Han Chinese may by now have turned the traditionally Uighur ethnicities into minorities in their historic lands.[28] They also seem to keep going in the same direction in Tibet.[29] Combined with centralized extreme surveillance and particularly harsh authoritarian measures, this continues to reduce their chances of ever successfully breaking away.

Xinjiang functions as a connection to Central Asia and beyond. It also delivers strategic depth by imposing logistical challenges because of its size and deserts. These give ample notice of many potential threats to the Chinese heartland.

As the home of the Silk Road that connects it to and through Kazakhstan, Xinjiang nowadays functions as a core piece of the "belt" in China's Belt and Road initiative (BRI). The goal of the BRI is to provide alternative routes for trade and energy supply, for example, to a Pakistani port and soon also an Iranian one built by India,[30] and beyond into the Middle East and Europe.[31] This is to make China independent of the foreign-controlled sea routes that we will look at next. It also is a reason the fall of Afghanistan to the Taliban strengthened China's position in this respect. Americans are no longer in the way, not even Indians. Instead, there are just China-friendly or neutral powers like Pakistan and Iran.

Tibet adds another crucial component to the security of the Chinese heartlands and population. Not only does its vastness provide a natural barrier for potential land attacks. Crucially, Tibet is the origin of China's major river systems. From a Chinese perspective, this makes its control essential to secure the domestic supply of water and food of over one billion people. For this reason, China would hardly ever seriously consider giving up control over Tibet. It therefore acts highly sensitive when it sees anybody challenging this and calling for Tibetan independence.

Although China, through Tibet, shares a long border with India, this border sees little trade. Both Asian giants contest it in at least two places, in India's northwest and northeast. Even just recently, this has led to several minor military clashes. But, there are no large-scale underlying disagreements involving either country's core interests. Geography clearly divides and protects both sides. In "Prisoners of Geography," Tim Marshall makes an excellent point when he calls the Himalaya and the Karakoram mountain ranges the "Great Wall of India."[32] These mountains are the geographical equivalent of "good fences make good neighbors." If the neighbors are not friendly to each other, it is their own choosing.

The mountain ranges get gradually lower on the border to Myanmar and Laos, but jungles ensure that they still act as an unassailable natural barrier. Like with most other parts of the Chinese land border, they do not facilitate large-scale trade or power projection. Even more important for geopolitics, China is militarily secure from land attacks from that direction.

On land, there are only two areas of serious concern, both when approaching the ocean. One is the relatively easy-to-traverse border to Vietnam. Despite its 100 million people and growing economy, though, Vietnam cannot pose a threat to core Chinese interests. It is more the other way around, simply because of the huge imbalance of population and economic numbers.

The corresponding latent concern in China's north is the long border to North Korea, which is located just about 800 km (500 miles) from Beijing. China's issue is not with North Korea directly but with what lies beyond: South Korea and Japan with their democratic political and market economic systems, capable militaries, and American troops. Despite Pyongyang's nuclear weapons, it poses little direct danger to China itself. Potential trouble is more likely to arise from North Korea's often erratic behavior. Pyongyang has been a constant source of instability and volatility.

This is the location of China's last major war, when U.S. troops approached its borders in the 1950s, causing a massive and overall successful Chinese counteraction. De facto, like then, the status quo works well for Beijing. North Korea partly functions as a buffer toward South Korea, and therefore also to the U.S. military stationed in the northern part of East Asia.

This leaves the seas to its east and southeast as China's major geographical challenge and concern. They add three close neighbors (South Korea, Japan, and the Philippines), one important island with a disputed status (Taiwan), and three other nations bordering the South China Sea (Indonesia, Malaysia, and Brunei). These maritime regions connect China to the world.

Even today, most physical trade between nations still is conducted by sea,[33] and this is true for China as well.[34] For China and for most of the rest of the world, trade usually fosters mutually beneficial interdependence. Globalized supply chains are a major source of wealth and know-how. In a similar way and mirroring other great powers like the United States or others before it, the seas are also the major option for China's navy to project military power globally.

Today, though, trade from and to all major coastal cities, including Guangzhou, Hong Kong, Xiamen, Shanghai, and Tianjin must either go through bottlenecks like the Strait of Malacca or pass by Japanese or Taiwan-controlled islands.

From China's perspective and for its self-image as a global superpower and regional hegemon, these sea routes reflect as much barriers as connecting roads. All access to the open seas, to the blue water oceans, has to go through potential choke points controlled by other countries. And all of them, for various reasons, are traditionally uneasy with their big neighbor or even outright opposed to a strong China.

And they all are also either formally allied or informally aligned with the United States.

The Island Chains

A closer look at what lies between the Chinese mainland and the blue oceans reveals islands, and lots of them. Geopolitical strategists use a combination of geographic characteristics and political context to assign labels to them, calling them "island chains."

The first string of islands is located mostly about 200 to 300 miles off China's coastline. This "First Island Chain" reaches from South Korea through Japan and the Ryukyu and Senkaku islands via Taiwan to the Philippines and then Malaysia. The "Second Island Chain" bypasses the Ryukyu Islands, Taiwan, and the Philippines. Instead, it links the Japanese Ogasawara and Volcano Islands to the U.S. Pacific territories of the Mariana Islands and Guam to Palau, and from there to New Guinea. Even further to the east, strategists pin the label of "Third Island Chain" on an assumed line reaching from the Aleuts Islands in Alaska through Hawaii to New Zealand.

All Chinese seaborne trade and its navy must sail through the First Island Chain, and therefore through waters controlled by foreign and often antagonistic nations. Here, I will not dive deep into the historical or political reasons. It is important, though, to keep in mind a straightforward Chinese perspective. The world's most populous country, its at least second biggest economy,[35] and the planet's second most powerful military has thousands of miles of coastline but cannot reach the blue oceans without risking interference by what it considers either a strategic competitor, rival, or even sworn enemy.

Chinese trade is subject to the interception of competing and even hostile countries, and its industrial centers are just hundreds of miles away from these opposing countries' military bases. And very active and with troops in all these countries, China finds the current global hegemon, the United States, with strategies and institutions in place that ever more explicitly try to counter and limit Chinese assertiveness. Western powers effectively check Chinese power not near Europe at the Strait of Gibraltar or near North America in the Caribbean but in China's "backyard." These are the coastal waters that the Chinese think their modern version of the Middle Kingdom should dominate, airspace that they should control, and seabeds that they should be able to explore for minerals and energy.

This perspective is independent of who rules China, of how they rule, and what their regional or global aspirations are. Any such aspiration seems limited through chains. They were golden chains when trade and the American-enforced rule of law stimulated commerce and economic growth. But increasingly the Chinese see them as iron ones. They perceive the First Island Chain as their version of an iron curtain, although inflicted by other powers instead of their own government.

Comparing this situation to the United States, it could hardly be more different. The U.S. is equivalent to a continent-sized island, nearly left to itself and bordered only by two closely allied and friendly smaller powers. From most of its major ports, the U.S. has uninhibited access to the blue oceans and the world, for trade or to project power militarily. Were the military and economic imbalance to U.S. neighbors not so lopsided, naval bottlenecks could exist for trade involving Gulf of Mexico ports like Houston, New Orleans, and Tampa. However, the United States looks at this body of water, and the Caribbean overall, as friendly waters and *its* natural backyard. During the past 100 years, this was only once seriously challenged in 1962. Then, Russia's actions and America's reaction brought the world to the brink of a World War during the Cuban Missile Crisis.

It will be helpful to keep this comparison in mind. When a strategic competitor showed up at American shores, Washington pulled out all the stops and put all options on the table, including total nuclear war.

Likewise, and appropriate or not, China considers the Yellow Sea, East China Sea and the South China Sea in the same way. And smack in the middle of the First Island Chain sits the Island of Taiwan. Its diplomatic status is ambiguous, to a large degree deliberately so. On the one side, about 180 nations, including all major ones, all but formally recognize Taiwan as being part of China. On the other side, many of the same countries, and particularly many U.S.-allied powers, practically support it as an independently administered and governed entity.

Does it surprise anybody if Chinese people see such deliberate ambiguity as a diplomatic game driven by self-serving economic, political, and military motivations? Particularly when, until the 1970s, these same nations seemed comfortable enough to accept the then-authoritarian regime of Taiwan as rightfully representing the government of the Chinese mainland as well? Then, they even had used that argument to give it a permanent seat on the United Nations security council.

For China, this diplomatic and political ambiguity provides the perfect opening to pursue what it considers necessary to take its destiny into its own hands. And so it vows to take control over Taiwan, one way or the other, and soon.

As I pointed out above, Taiwan is located roughly 160 km (100 miles) off the Chinese coast, which is about as far away as Cuba is from Florida. It is only about one third the size of Cuba (33,000 km^2) but with almost 25 million people twice as populous and with a trillion-dollar economy almost ten times Cuba's size. It has claims on islands currently controlled by Japan and others contested with the People's Republic of China and the Philippines.

Relevant in case of a war, Taiwan controls and administers two islands directly next to the Chinese mainland, Quemoy and Matsu, just a couple of miles offshore. The Kuomintang had used them in the past to amass military forces that threatened the invasion of the Chinese mainland in 1958.

When one looks at a map to see the exact shape and flow according to the above definition of the First Island Chain, one noticeable irregularity becomes obvious. To include Taiwan as part of this chain, one must swing much closer to the Chinese coastline, and then back out again. It would be equally appropriate to define the First Island Chain as going straight south from Okinawa toward Luzon in the Philippines. As a defensive line for American allies in the 2020s, this would probably make as much sense as the current definition. But it would exclude Taiwan and make Formosa and other islands look more objectively like a natural part of China. This is exactly as China sees it, and as Taiwan and the U.S. did until just a few decades ago.

It is just one ambiguity of many that reveals a lack of objectivity of geopolitical "truisms." Even with maps and definitions, we quickly descend into politics.

But it may not be all about Taiwan.

Geopolitical Aspirations

China's goal is to break out of the constraints imposed by the U.S. alliance system, and be able to freely project force beyond its immediate neighborhood. It wants to protect its interests according to its own interpretation and without limitations. This means full control over its own supplies of energy, food, raw materials, supply chains, and ideally also dominating the rules in its neighborhood. China also wants the ability to mess up whoever interferes with these strategic goals.

If that were not the case, why would China otherwise build a modern navy that rivals the U.S. Navy, and even exceeds it in numbers? As Kevin Rudd lays out in fair detail in his 2022 "An Avoidable War," President Xi is pointing the PRC in exactly this direction. The officially published "China's Military Strategy" defines "the centrality of the navy and the maritime domain in China's overall strategy."[36] Clearly, the oceans, maritime power, and therefore dominance and control over what China sees as its own First Island Chain (reaching into the South China Sea) are highly relevant geopolitical priorities for China.

During the next decade, China's People's Liberation Army Navy (PLAN) will likely become the largest modern blue water fleet in the world by any definition. This would then not just be in numbers of ships, which already is the case today,[37] but also in tonnage. Currently, all these ships are based in its extended coastal waters. Considering the closeness to missile bases and airfields on the mainland, this creates an ever more lopsided situation between the PRC and Taiwan. Even if the U.S. continues to keep between 50% and 60% of its naval force focusing on the Western Pacific,[38] as it does today, it may barely be able to match Chinese naval might inside the *Second* Island Chain, even together with all American allies.[39] Many analysts think that this is already the situation today.

China does not want over 500 ocean-going ships, including several aircraft carrier groups, bottled up in its coastal waters. It will not live with that any more than the U.S. or Great Britain would. Why would the world's biggest manufacturing and trading power build the biggest fleet in the world if it cannot access the blue oceans without having to go through areas controlled by strong opposing powers? Note that, because of geography, this is not a challenge for many other great nations, like the United Kingdom, India, even Japan. It is similar, though, to how Germany perceived itself bottled up before World War I, and Russia has been for centuries and still is today. Great powers, and even more so rapidly growing ones, do not take these things lightly.

And so, the Chinese goal is to shatter the constraints of the First Island Chain and gain unmitigated access to the deep seas.

How Technology Transforms Sea Power in the 2020s

All this reflects a traditional view of geopolitics and is still highly relevant today. But it also is increasingly being superseded by technology. Geography, the larger geopolitical landscape, and above all technological advances make it unlikely that China can achieve access to blue oceans in a fully "unmitigated" way similar to those of some other global powers in the past. South Korea, Japan, and the Philippines are not going to physically move out of the way. Their islands, and South Korea's peninsula, will always be close to Chinese trade routes and the pathways used by the PLAN. Breaking through the First Island Chain can therefore give China better access to blue oceans, but not undeniably so.

We can even challenge whether, starting soon, *any* ever will be able to "freely" traverse oceans. Not even the United States has this anymore to the same degree it had in the past. All nations are losing such a level of "freedom of navigation." Somebody always can track, stop, and destroy navies.

This is an unprecedented change. Even as late as during World War II, both the starting knell with the surprise attack on Pearl Harbor and the first Japanese defeat during the Battle of Midway were assisted by the enemies not having a clear picture of each other's positions. But in the 2020s, satellite surveillance and new technologies make it ever more unlikely that ships can hide anywhere, and definitely not large surface vessels. And then, despite sailing and therefore constantly changing their location, they become targets for the ever expanding and spreading ability to hit ships. Today, one can, at least in theory, achieve this with advanced missile systems, submarines, smart robotic buoys, and other even more asymmetric anti-access weapon systems. These include flying or swimming swarms of drones and space-based kinetic payloads.

Friend and foe alike can see the U.S. Navy leaving its home ports, and ever more closely track each ship's exact positions and movements. Soon, this may extend to American submarines as well, as artificial intelligence can analyze wave patterns and find submarines well beyond a depth of 1,000 feet.[40] The U.S.' most advanced submarines are still safe in many waters because they can dive deeper than even the most modern detection technologies can penetrate. But for how long? And will we know when it is over and Chinese AI systems succeeded with tracing them wherever they are?[41]

The above developments have not changed the traditional view of geopolitics that still today dominates the great power's thoughts and strategies. There are strong political, cultural, and historical reasons for China wanting to control Taiwan, as there are comparable or even stronger reasons to resist. The fact is, though, that in the traditional view of geopolitics, and *only* in such a traditional view, the sole chance for China to break out of the First Island Chain is by controlling Taiwan and its adjacent islands. To consider the threat of impending great power war, but also to seek solutions preventing it, we must keep this point in mind.

Other pathways would not work. Even if China would control the whole South China Sea and its islands, the Strait of Malacca would still be too narrow and easy to block. So are the shipping routes through the islands of Japan. Navigating through the thousands of islands of the American treaty-ally Philippines, or Australia's friendly neighbor Indonesia, is also unrealistic and out of the question. There are too many bottlenecks and choke points stacked on top of each other. The only alternative way for China to be substantially in control of its physical trade and ability to project military force is by going through a large enough gap between Japan's main islands and the Philippines.

This means Taiwan.

We do not need to accept this traditional geopolitical calculus and hopefully can shift it by pointing at implications of the before-mentioned technological shifts. But as long as we consider conventional viewpoints as crucial, as both the U.S. and China currently do, we have few logical arguments to make China's perceived geopolitical issues go away. We could ask China to stop its naval build-out, refrain from asserting itself, and abolish its goal of rivaling the U.S. and projecting force globally. This would imply it accepting the continued hegemony of America or at least of the American-led alliance system, even in Asia.

Fat chance.

Geopolitics of the Future

Unfortunately, the just described way of looking at geography means that now, China and the United States view the world through the rearview mirror. Much of their strategic thinking is dominated by "island chains," "access to blue waters," "freedom of navigation," the supply of fossil energy, the flow of supply chains since the 2000s, and the commerce of physical products via the oceans. They consider modernized versions of old technologies and military capabilities even though nowadays these hardly even work against vastly inferior forces. They think about the capabilities on land, at sea, and in the air.

Of course, all these are important. But they are no longer the solely dominant frameworks enabling and driving our world today and will become ever less important in the near future. The old strategies all focused on the old commons of the sea and air as the main pathways to ship most goods. But the 2020s and beyond are about the new commons, the spatial commons: space and the digital.

To protect our homelands, we must protect them from space attacks. And we must secure the digital world to ensure the supplies of food, water, electricity and energy for our people. Modern societies need the ability to communicate, drive cars, use hospitals and emergency services, fabricate products, and keep a society running.

Although the volume of most global trade is shipped via the oceans, this hides the actual picture. Already today, the largest component of the commercial value traded along physical supply chains that navies can protect is, in reality, digital.

For example, most value and cost of a modern car comes from its electronics and not from its physical components. The same is true for many of these physical components themselves, like the car's batteries or sensors. Likewise, most products are designed digitally. We can send these digital designs to machines that automate their fabrication through robots - anywhere. These robots' capabilities require no training but simply the installation of computer software. Once fabricated, manufacturers link ever more products to virtual duplicates called "digital twins." They can then maintain and remotely support them even from other continents, by service personnel wearing augmented reality (AR) goggles.

All this is already functional today and spreading rapidly. It is transforming work, manufacturing, and support. Increasingly, value-generating programming, design, and maintenance activities are performed by globally distributed teams using virtual platforms and high-speed communications networks. They leverage data collected through billions of sensors, subsequently optimized through AI and advanced computing.

Simplistically expressed, given sufficient raw materials, goods and products can in theory be manufactured anywhere. And at a logarithmically increasing rate, exactly this happens today in practice.[42]

What this shift needs and is based on, though, is functioning, freely accessible, and exponentially increasing computing power and communications as provided through space and ground-based networks. It is why I call these the new commons. Most of today's commerce and of our manufacturing and services need and use such digital capabilities. Likewise, this then also is where traditional and new forms of warfare will be conducted and where actual harm can be caused, including physical destruction and death.

All this changes the meaning of geography and physical spaces.

The old world of projecting power and fighting wars focused on land, sea, and air. For thousands of years, humans fought wars on land. Many times, oceans and rivers also played important roles. But even in early 19th century Europe, not the navies but infantry and cavalry eventually ended Napoleon's wars on land.

Particularly since the enlightenment and its resulting technological advances for naval capabilities, though, the power of navies grew. It enabled the flow of people and their supplies, and of high-value resources like spices or gold. This soon extended to bulk food, raw materials, and sources of energy. Subsequently, the strongest naval powers dominated most of our planet. Ships became more powerful, using sails, then coal, petroleum, and some of them even nuclear power.

Starting with World War I, air power entered the realm of warfare as well. In World War II, it played a crucial role in fighting armies and civilian populations, up to when airplanes delivered the most potent of all weapons, nuclear bombs.

During the Second World War, vastly enhanced capabilities of land, sea, and air were the most effective way to conquer geography and control countries and people. And even today, this is on its most basic level still the focus of military strategies and policies.

Recently, these strategies have started to consider ever more digital threats as well. The U.S. Department of Defense (DOD) also leverages space-based communication, GPS for navigation, and even some civilian communication systems. The DOD understands that it must protect these. But at its core, reflected in military budgets and weapons systems, little has changed compared to World War II. Its acquisitions still seem mostly about better versions of the old weaponry. And about similar strategies.

Some people call that "fighting the last war."

In the 2020s, however, the old definitions of threats no longer reflect the full reality. Arguably, they are not even this reality's most important part. Even if we exclude nuclear bombs, traditional weapons no longer are the major challenges for our citizens and our societies.

Space and the digital realm are transforming the meaning of geography. This applies both to warfare among militaries in the traditional sense, and to the way it brings war home to each other's societies.

Space-based weapons can take out our communication satellites and therefore render useless many of our capabilities abroad. Enemy satellites can see our ships and, to some degree, even our submarines. They can therefore find and attack them wherever on Earth they are. But most importantly, they can directly attack our homeland using intercontinental missiles, space-based kinetic payloads, long-distance semi-autonomous drones, and electromagnetic pulses (EMPs).

Digital weapons and acts of sabotage can directly damage and wipe out the critical infrastructure of our homeland even without such weapons. These capabilities go far beyond conventional cyberattacks. Their potential effects can extend to little less than the de facto destruction of how our modern society works.

It does not render geography useless. For many countries, supplies of bulk food, raw materials, and sources of energy may even become *more* critical. But not intermediate products and the same kinds of complex supply chains we were used to during the past decades. We must develop a completely different perspective on how to use and secure those spaces that matter for the survival of our armed forces and our society.

All such actions are to protect and empower people. Therefore, we will look at the world's populations next.

4

Demography

The Many and the Old

"The most populous country in the world." Chances are, this is one of the first factoids that comes to people's minds when thinking about China. It is not just the number itself, though, but how much larger it is than those of all other countries but India. The United States, as today's number three in the world, has a "mere" 350 million people, compared to China's 1.4 billion. This quickly can be perceived as threatening, particularly while the Chinese population becomes ever wealthier. And even in the early 2020s, and despite continuing mass immigration and population growth, the U.S. numbers seem to fall behind ever further.

This part of the picture, though, is about to change and then raises a distinct set of questions. Is it possible that this vast difference disappears during the 21st century? Could the population numbers of the U.S. and of China converge, and America and China become similarly populous? And when we then compare their numbers to other countries, what could this mean for the Chinese-American relationship?

At a first glance such thoughts seem preposterous, considering today's numbers. Diving deeper into the details, we quickly realize, though, that such a development is not only possible but also a helpful thought experiment. It reflects how volatile and impactful demographic developments can be.

Today, China benefits from the current demographic imbalance. The first decades of China's rapid economic expansion since around 1980 leveraged its vast numbers of young and cheap manual labor. The PRC pursued a deliberate strategy to expose its society to the world economy and capitalist market dynamics. This succeeded by injecting the country, and its people, into nearly all global supply chains and therefore also into the modern world.

A look at past and current global trends lets us better appreciate variations of future trajectories. We then see that the large population increases that we witnessed in the past two centuries were historically unique. More recently, they also seem to abate significantly. And it is unclear where we are heading from here.

Like the size of economies, the world's population changed little during most of the past two millennia. Around 2000 years ago, estimates of the world's population stretched from 200 million to 400 million. By 1650, they hovered around 500 million. These numbers were subject to swings because of diseases, famines, and war. But for centuries, they were overall stable. During any person's lifetime, population growth was nearly imperceptible. Incremental growth rates stayed far below 1%.

And then came the European enlightenment, which triggered the industrial revolution, the availability of massive amounts of energy, and medical and other scientific advances. Everything changed after that. Population growth rates increased far beyond 2%, before flattening out. 2% means a doubling of a country's population within 35 years. This was noticeable even to the average person. Population numbers doubled and tripled during a typical human lifespan.

And then, in almost all countries, the numbers stalled and are now projected to shrink soon if they are not already doing so today.

These swings can be confounding. Population numbers should be straightforward to forecast. It is mostly a matter of taking the total number of women in childbearing age, calculating the fertility rate, and extrapolating it into the future. As we will see, though, shifts and swings are not just larger than ever before. They also can go in both directions and diverge between nations.

This has profound implications for today's geopolitical system, for world peace, and for the global system during the rest of the 21st century. And China's and America's roles in this are more dynamic than most seem to recognize.

A look at the demographic picture in more detail reveals some geo-strategically highly relevant facts.

2030 - The Demographic Inflection Point[43]

At the beginning of the 2020s, China's population of about 1.4 billion means that it is over four times as populous as the United States. Current trends show that India will surpass it soon, probably within the next five years.[44] This will be shortly after China reaches an inflection point and its population number has started to drop.

By the middle of the century, and after shrinking for a couple of decades, China's population is likely going to be far below today's number. And then the current demographic picture really will shift, with widely diverging estimates and projections. While the United Nations Population Division predicts a subsequent decline to still just over 1 billion by the end of the century, this would require an increase of today's fertility rate from what this forecast assumes to be 1.59 in 2020 to then 1.81 in 2100. At least so far, though, there is no sign of this happening.

Fertility rates have a significant impact. Were China's rate to settle somewhere between the current number and the U.N.'s prediction of 1.81, a model of Vienna-based Wolfgang Lutz and colleagues forecasts 754 million Chinese in 2100.[45] Assuming a rate more similar to the somewhere between 1.4 and 1.5 that some other large countries experience, models of both Lutz and of the U.N. would see the Chinese population drop to between 612 million and 643 million by 2100.

When this seems too pessimistic, it may be the opposite.

Official Chinese statistics pin the current fertility number even lower, at around 1.2. Like in India, a disproportional number of these are boys - about 20% more than girls.[46] By the end of the century, China's population would then be somewhere around half a billion people. This would be roughly a third of today, and could end up being lower than the population of the U.S.

It is tempting to attribute these falling numbers to China's one-child policy. If this were correct, the recent lifting of this policy should change the trend. But there is little evidence supporting this interpretation. Other countries, including Taiwan, did not have a one-child policy and their fertility rates dropped even below China's current numbers. Taiwan, Japan, Iran, or European countries are just some examples of what has become a global trend. Female education seems to be the primary driver for these rates, not governmental policies, cultural values, religion, or the level of economic development.[47]

Even Middle Eastern and African countries are following the same trajectory, although many only recently have begun the typical precipitous fall of births.

Only one large country seems to be prepared to escape this long-term trend, representing a special *kind* of countries. Nations with strong immigration are predicted to continue increasing their population.

Like the United States of America.

The United States: Fertility and Immigration

The United States is likely going to continue growing at least through the middle of the century because of its comparably higher birthrates, and above all because of immigration. It will then have added 100 million people and reach 458 million.[48] From then on, the trend is unclear and could, according to the UN, approach up to 600 million by 2100, or fall back to approximately today's level.

Like with China's numbers above, this conservatively assumes an only incremental improvement on today's medical capabilities and healthcare systems, including the ability to battle diseases or even potential pandemics. Below I will explain why some of these assumptions may be incorrect, though.

Policies are a big unknown, particularly those about immigration. For example, considering America's history and the continuing inherent natural and economic wealth of North America, a deliberate swing toward even more immigration is possible. Even a still comparably moderate and manageable annual expansion of about 1% could cause a doubling of America's population within this century. This would be the impact of immigration alone.

Most other countries would have a much harder time even considering such an acceleration. But for the United States, there even is a historic precedent for this. During the first 100 years of its existence, and to a large degree driven by the cumulative effect of immigration,[49] the U.S. doubled its population about every 30 years.[50] Roughly half of this growth can be attributed to immigration.

The big unknown, therefore, is America's immigration policy and whether and to what degree the country continues its appeal for global talent. A popular recent book by entrepreneur and best-selling author Matthew Yglesias explored the implication of such a world. The book's title "One Billion Americans" is a contrarian exploration of a strategy that would be relatively simple to implement. Almost all it would take would be to replicate a milder version of America's 19th century experience.[51]

The result would dramatically alter and even upend the demographic picture in the world. Demographically, China would turn into an aging power, and possibly a stagnant one, while the U.S. would become comparably young, dynamic, and therefore disruptive.

So far, so good. Neither of these, however, may be the most important and impactful demographic trend on Earth.

Europe, Japan, India, Africa, and Others

Other countries also play into the larger geopolitical picture. Outside North America, the populations of today's largest and most powerful economies in Europe and East Asia are all more likely going to stagnate or shrink.

Europe's trends depend on how it deals with immigration. Without continued immigration, its societies will contract. This will shrink its overall economic power and reduce its capability to continue its technological and innovative dynamism.

The populations of China's East Asian neighbors, Japan and South Korea already are dropping and aging today. Since China's population will roughly remain stagnant for a few more years, this will continue to extend the current economic and military power imbalance in the region in China's favor. After about 2030, and independent of America's decisions, China will then join its two other Asian nations and two other world regions whose population numbers already have started to stagnate: Europe and Latin America.

South Asia and Africa, however, will continue to experience growth for a longer time. Should they continue their recent economic and technological advances, this may trigger another geopolitical and geo-economic shift in global power relations.

India is already in the middle of this, while Africa's challenges grow almost as fast as its advances. This vast continent is too diverse and decentralized to enter the world stage as one block, even compared to the also highly diverse Indian subcontinent. Africa's ability to translate its population growth into prosperity and more power is difficult. To the largest degree, it depends on whether its countries can leverage digital technologies and get integrated into a larger, open, and liberal economic system. If so, Africans will get the chance to leverage the entrepreneurial and innovative potential of their young population. This could turn the continent into a key contributor to economic growth and strengthen a free world. We will explore this a bit more in Part III of this book.

Nigeria is emblematic for what is happening in Africa. Continuing its currently reported growth trajectory, this West African country may within decades slightly outgrow today's population of the U.S., to beyond 400 million. It is open where the country goes from there.

But the primary driver to change the geopolitical balance in Asia will come from India. When this nation passes China as the most populous country on Earth, and if it continues its current profound economic growth rates, it may become an important wild card in geopolitics.

In much of the Western world, this flies below the radar, like China about two or three decades ago. India's ambitions are regional first and global second, mirroring China's during its ascent (and America's). But as India's role increases, continued clashes with China are pre-programmed, and could very well change China's geopolitical rationale.

Most of the conflicts between China and India are likely to be about predominantly symbolic matters or geo-economic ones. Among the first category are the border conflicts in the northeast and northwest of India. Except for control over the origins of rivers, neither of them could materially alter the strategic balance. This makes them mostly symbolic. But there are other signs of rivalry. Examples are the expansion of the Indian space program, and Indian ambitions to build indigenous aircraft carriers.

For these reasons, the world's continued focus on China as the "new kid on the block" in great power politics is behind the curve. It is reactive instead of prospective. Policy makers and public perception respond to *past* trend lines, not prepare for those of the near or far future.

We can summarize the above as follows. Current trends and probabilities for the next few decades suggest the populations of China, Europe and Latin America to stagnate and contract. East Asia and even much of Southeast Asia will follow suit. The regions with strong population growth will be India and Africa. North America will grow, but hardly as fast as India and above all Africa.

The tipping point will be at around 2030. After that, the most dynamic world regions are likely going to be India, Africa, and North America. To continue its current path of creating wealth for its people, China may find itself in a role more similar to Japan and Germany of the early 21st century rather than the early 20th century. For India it may be the opposite.

And demography and related economic growth may not play out as deterministically as, for example, Singapore's Kishore Mahbubani keeps stating.[52] The world is not binary, and the term "arc of history" can be misleading. There is less of a pendulum that swings back to return Asian powers to their position at the top of the world's food chain. Rather, the analogy of "quantum states" makes more sense to hint at the never-ending variety of paths that history can take.[53] As the above numbers and projections show, this even applies to something as seemingly straightforward as demography.

Geopolitical and Long-Term Consequences

India's population surpassing China while keeping a young population will have a big psychological effect. For the first time in modern history, China will not be the most populous country on Earth. And while China rapidly grays, India is likely going to churn out millions more engineers and professionals than its northeastern neighbor.

The Next Generation

The combined effect for East Asia will be a desire of China to quickly establish a favorable geopolitical environment. In his interpretation of trajectories, China would have to act *soon* to improve its geopolitical power position, or it may be less able to do so afterwards.

Let's unpack this and apply it to our main topic. Demographic trends motivate China to establish by around 2030 as much control over its immediate neighborhood as possible. It will intend to secure the reliable supply of raw materials and energy however it can. This includes control over adjacent islands and securing core supply routes. And indeed, it is exactly what China has started with its Belt and Road Initiative and the Maritime Silk Road. Politically and militarily, this is explicitly China's goal and the main reason for expanding its navy. The PRC wants to establish control over the South China Sea, and soon grab and incorporate Taiwan and its adjacent islands, by any means necessary. See President Xi's quote at the very beginning of this book.

But another demographic tipping point may change these dynamics. Within the next few years, again by around 2030, the age composition of the Chinese population will rapidly reduce its workforce and increase its number of retirees.

Most of the world's pension systems are still administered by governments, reflecting century-old solutions to accommodate industrialization. When a growing share of the population must fall back on such governmental support, this limits budgets for other purposes, including infrastructure and the military.

China already ages faster than the U.S. and will continue to do so. In 2050, the U.S. median age is predicted to be 41 years, while China's will be 46, and Japan's, South Korea's and Germany's all higher than 50.[54]

The effect is unlikely going to be a quickly noticeable shift away from China's current advantage in science, technology, engineering, and mathematics (STEM) education and national research and development (R&D) budgets.[55] For several more decades, the U.S. will only be able to graduate a fraction of the engineers that China does. However, aging is probably going to indirectly slow down Chinese growth numbers. Despite a huge savings rate, particularly compared to the U.S.,[56] the aging of China's population is going to become an economic and political burden. China's retirement systems will create additional pressure for its government because it still has hundreds of millions of people that have not yet achieved comfortable middle-class incomes and lives.

Already today, the PRC's levels of debt are tremendous. They are still higher than in the U.S., although the U.S. has recently been catching up. In China, the government strongly regulates most options for saving for retirement. It limits investment in international markets. What remains is often just investment in real estate and in state-owned or controlled businesses, which as a result creates bubbles.

Once China's population stabilizes and then contracts, it may force some of its geopolitical calculus to change. The general trend will also lessen pressure from a young and restless working population. Instead, it may focus attention and its budgets more on addressing domestic issues of retirees rather than military or foreign policy adventures.

Beyond 2050: Technology-Driven Evolutions and Revolutions

What then is going to happen to demographics after 2030, and particularly beyond 2050?

Although the U.N. and others make predictions until at least the end of the century, these have severe shortcomings. Part of the reason is that so much hinges on the cumulative effect of fertility and mortality numbers. And there, trends can change quickly. Like fertility rates spiked during industrialization from 2 to 3, 4, and even 5, they dropped precipitously when the information age arrived, from between 4 and 5 to below 2, often within just a few decades. What is to say that they can't go back up above 2 within a decade or two?

In this respect, two developments may have profound effects. Neither requires a big change in human behavior because both involve technological advances.

The first of these is about mortality rates. During industrialization, much of the initial population explosion was driven by medical advances. Falling early child mortality combined with improved hygiene and better nutrition, while the number of children born stayed high. Only later, and particularly when women got educated, fertility dropped, and population growth slowed and reversed. A remarkably positive result of this process has been that life expectancies of nearly all countries on our planet are converging.

Even people living in most poor countries live longer than 60. The growth numbers for the most advanced countries have slowed down but continue to edge toward some number between 80 and 85. In mathematical terms, this number is being approached asymptotically. This means that further advances in average longevity take ever smaller steps and therefore may not extend far beyond about 85 years. It seems that this is how old humans can become before they die, on average.

Although, this probably is wrong. And it leads to an even bigger wildcard.

Evidence keeps piling up that aging is a disease and that we can defeat it and therefore substantially lengthen our lifespan. [57] Simplistically expressed, aging is a process of cumulative environmental effects compounding until eventually our cells no longer can sufficiently repair themselves. This weakens our bodies and triggers ever more diseases. We keep battling one after the other, then multiple of them together. Eventually, our bodies can no longer cope, and we die.

Modern science has identified at least half a dozen pathways to stop this process.[58] It is a rapidly expanding and very promising area of science and biotech advancement.

A growing number of reputable scientists suggest that our eventual lifespan may be between 120 and 150 years.[59] Mostly healthy years at that. Scientifically, there is little reason to assume that a 100-year-old could not be as healthy or productive as a 60-year-old or even a 40-year-old. Obviously, if people would live up to twice as long as today, even a very low fertility rate would then result in a further expansion of the population.

This is not all, though.

Technology can even overcome the limits of low fertility rates. Researchers of the California-based Buck Institute are making breakthroughs that extend female reproductive longevity.[60] The goal is to extend the time span during which women can get children. If we combine this with a longer productive and healthy lifespan, the fertility rate may improve. Getting kids in the 50s or 60s, midlife after being established in life, may seem an appealing option. Career could come first, children second, and a second career afterwards. The options multiply, and maybe the overall number of kids also.

Even more futuristic is an indirectly related scientific achievement. Today already, science has established that mammals can be genetically designed and then grown in artificial wombs. And, for this purpose, humans are just mammals, too. As of the writing of this book, Austin-based Colossal Biosciences seems mere months away from achieving exactly this for re-creating the woolly mammoth.[61]

Such a scientific achievement opens the door to some unprecedented capabilities. If we also add today's ability to perform changes to our genetic code, a society could then artificially increase its population numbers with more healthy and capable people, with no need for pregnancy or childbirth.

Here is not the right place to dive into more details, nor discuss the ethical consequences. We just should note that none of the just-mentioned technologies falls into the realm of science fiction. Rather, it would only require piecing together and using scientific capabilities already available today. For any advanced society, particularly one that is unscrupulous enough, biotechnology can open doors to overcome what in the past seemed the most basic limitations of biology and therefore also demography.

Geopolitical Consequences

With all this above in mind, what should we then do when analyzing geopolitics and developing future-oriented policies and strategies? And what does this mean for the possibility of a war between China and the United States?

Above all, we must keep an open mind about demographics. Let's not assume that current trends will continue forever, even long-term ones. An example of the United Nation's population graph about China reveals the challenge.[62] The population forecast looks like an inverse funnel, similar to scientific models of climate change, or the Weather Channel's likely path of a hurricane or cyclone. The further we go into the future, the larger the numerical spread of various scenarios. Regarding China, their median numbers suggest a drop in population numbers from about 1.44 billion in 2020 to roughly 1.05 billion in 2100. The full bandwidth, though, ranges from below 700 million to almost 1.6 billion, with statistically defendable outliers falling even to 500 million.

We have a good enough idea of where the next few decades likely are going to lead. Therefore, we can and must prepare for it. For the longer term, things are much more difficult.

For the competition between China and the U.S., we must note that China is about to extend its demographic "lead" only for a few more years. As an older and still significantly poorer society than the United States, the PRC then will shrink while the population of an immigrant-friendly America should continue to grow. Within a few decades and the lifetime of many of those currently already born, much or most of the difference in the population between China and the U.S. is likely going to shrink. The gap could even completely disappear.

For sure is only that we must consider the current numbers and trends with great care. Even over just a couple of decades, substantial changes of trends are possible. These then can have significant implications for geopolitical power, questions of war and peace, and therefore also the possible or actual implications of a war between China and the U.S. However, because of technological advances, predictions beyond the middle of the century are extremely difficult to make.

In the meantime, population numbers of South Asia and Africa will have far surpassed China's. This suggests that within a decade or two the rivalry between China and the U.S. may not be the most important one on the planet anymore. What seems like an existential threat and cataclysmic conflict is more likely little more than a temporary blip. Barring any major policy changes, we can expect the dynamics and perception to change because of other world regions' and both China's and America's domestic political developments.

A medium-term or long-term view of the U.S.-China rivalry should consider these shifting dynamics. The next decade is going to be a rollercoaster. A realistic and noble goal could, therefore, be preventing a large-scale military conflict between China and the United States at nearly any cost possible. We can set in motion a long-term oriented process acceptable to both sides while letting things settle for another 10 or 15 years. By buying time, both societies can mature their relationship and adjust their priorities to the coming technology-driven upheavals and transformation. These are much more serious challenges than even control over a large island, or many smaller ones, or the places that container ships, fishermen, and airplanes can freely traverse.

All this, though, depends on several additional factors. The most important ones are economic. And therefore, we must take a deeper look at economics next.

5

Economy

Capitalism Still Rules

The "rule of 70" lets us quickly estimate how many years of consistent growth it takes for a number to double.[63] Dividing 70 by the annual growth rate in percent will give us this amount of years. For example, if a country's economy grows by 2%, it will take about 35 years to double its size (70 divided by 2). Would the country grow by 10% (70 divided by 10), its economy would double in seven years. Within the same 35 years, the faster growth rate would cause five doublings. With 2% annual growth, the country would have grown two-fold, with 10% growth, the same country would have grown no less than **32 times**. That difference is a factor of 16!

This is basic math. And it is roughly comparable to what happened between the U.S. and China.

During the past 35 years, China's economy has grown 47 times in nominal dollars, as used by the World Bank, or 80 times in purchasing power parity, or PPP, as used by the International

Monetary Fund (IMF). The U.S. economy grew less than fivefold, or between one tenth and one sixteenth of the Chinese number.

As it goes with exponential numbers, even compounding percentages, they initially appear deceptively small, but suddenly they balloon and seem overwhelming. This happened when, as if out of nowhere, China overtook the U.S. economy. And because of this perceived suddenness, many feel threatened, particularly in the U.S.

Big Australia, Tiny Australia, and Small Japan

A generation ago, China's then-leader Deng said that China should "bide its time."[64] The PRC should first build wealth and underlying capabilities before aspiring to shatter limitations for the country's international actions. Following this advice, since the 1980s and 1990s, the country's strategy has been to accumulate overwhelming economic power and translate it into regional military superiority. Only recently, the PRC slowly and carefully began ripping up and reshaping East Asia's and, to a lesser degree, Southeast Asia's geopolitical structure to serve China's persistent desire to become even more prosperous, powerful, and secure.

If China's higher economic growth rates continue, its small neighbors will hardly be able to stand up to it. Korea, Japan, the Philippines, Vietnam, even Malaysia and Indonesia, could become mere bumps on the road toward China's regional hegemony and dominance.

At a first glance, it may sound bizarre to call Japan a mere "bump in the road" for China's might. But what may have sounded strange and upside down as late as in 1990 is today becoming a reality. It was not even two decades ago when China's GDP surpassed Japan's.[65] But today the PRC's economy already is far more than double the size of its East Asian neighbor.[66] And military power is following suit.

China's ambitions may sound even less like a fantasy when considering two facts: According to World Bank statistics, in 2018, China's economic *growth* was larger than the GDP of the whole Australian economy. This is the same consistently and rapidly growing Australia which just 30 years earlier had an economy larger (!) than China's. The almost exactly same trend is on the way between Japan and China. As recently as in 2008, China's economy was smaller than Japan's. But according to a forecast of a leading conservative Japanese think tank, by the year 2060 the GDP of Japan may have fallen to around 10% of the Chinese.[67]

Yes, this is not a typo. It is **ten percent**.

The implications are dramatic. They will upend traditional power relationships, regionally and globally.

China will not even have to directly threaten Japan militarily to get its way in most fields that it considers important. Even if old enemies Japan and South Korea combined forces, they would only have a minimal ability to project any kind of economic or military power, when compared to China. The best they can realistically achieve is to keep their national integrity, cultural identity, and sovereignty over the core geographical area of their nations.

Like with almost every other Asian country (including Taiwan), China is already today Japan's and South Korea's biggest trading partner. Unless they find a strong alternative in the global system that goes beyond the U.S. current abilities and vision, both nations will naturally fall into the Chinese sphere of influence and de facto economic hegemony. Such an alternative does not exist at this point and the current system got them into the dilemma that they are facing right now.

The title of a popular book about business strategy and personal development points out that "What Got You Here Won't Get You There."[68] This statement applies to international relations as well. In Part III of this book, I will elaborate on what to do to get us "there."

In summary, it will require a new kind of framework, a technological platform for digital commerce, supported by multiple large countries. This would establish a future-oriented global extension and evolution of today's Internet and base it on liberal principles. It must be future-oriented to account for underlying shifts of the monetary system, the digitization of supply chains and the flow of data, and divergent regional economic growth rates. We must not protect the oceans above all, but the new commons of space and the digital realm. These determine future economic and military might.

In the meantime, and unless they recognize progress in this direction, China's neighbors would fare better to find some arrangement with China that eliminates sources of conflict and draws hard lines while they still can.

From Below the Radar to the Top

With us humans being used to slow and linear growth makes it easy to get consumed by zero-sum thinking. We assume that significant shifts favoring one side only are possible when the other side loses. Or the other way around: when others win, we lose. It is this thinking about our economic relationship with China that makes us feel threatened. A logical reaction would then require us to prevent the other country's rise, stop its continued accumulation of wealth, or at least slow down its economic growth to a rate below ours.

This would be wrong for several reasons. For one, it would mean prohibiting the development of people in another society. Doing so would be the antithesis of what America stands for, which is advocating for everybody's "pursuit of happiness." Even more important is that zero-sum thinking reflects a mindset of scarcity instead of abundance. In reality, (still) exponentially increasing capabilities of digital technologies open an abundance of opportunities, as I will explain in more detail in Chapter 7.

Such advances create the chance for everybody to accumulate wealth and prosperity. To maintain our relative power position while others become richer, all we would need to do is grow faster than them. During the past four decades, though, the opposite happened.

China had economic success following state-sponsored strategies that deliberately unleashed transformative raw capitalist forces. It enabled the PRC to build a technology and manufacturing base today matched only by the United States.[69] During the two decades of American focus on domestic battles and the Middle East with its vague concept of a global "war on terror," the gap has all but closed.

According to the World Bank, China's economic output has grown from below 12% of the U.S. in the year 2000 to beyond 70% in 2020. This is in nominal terms, meaning it even distorts the picture in America's favor. According to purchasing power parity (PPP), the IMF and America's Central Intelligence Agency (CIA) say China passed the U.S. in 2017.[70] And for over a decade already, starting in 2010, the World Bank has considered China the globally largest manufacturer. Although U.S. policies since 2017 started to slightly reduce the Chinese lead, its manufacturing output is still over 50% bigger than the American one.[71]

All this does not bode well. The old hegemon, the United States of America, has a hard time adjusting to this new reality. The U.S.

often acts as if it were ignorant of its diminished stature and increasing economic and technological vulnerabilities. We close our eyes and prefer comforting fiction over reality, and history over the present. This then also affects how we assess our military capabilities. The challenger, the People's Republic of China, brims with overconfidence because of its obvious advances or outright superiority in many fields we traditionally consider when assessing power.

Emboldened by economic success, China may regard the risk of war with the U.S. as manageable or even minimal and the possibility of losing as even smaller. After all, not just the absolute size of economies count. Many U.S. companies depend on Chinese money and markets, U.S. consumers want Chinese-made products, and U.S. and other Western manufacturers rely on critical Chinese contributions to their supply chains. Comparing the superior numbers and capabilities of its military in the regional theaters may stiffen Chinese resolve even further.

And even if, or when, some of its economic bubbles burst and weaknesses surface, this does not help much either. When overexposure in real estate and debt potentially endangers the value of insufficiently secured pensions, it even may push Chinese leaders to accept limited war as a welcome distraction.

China has an economy equal to America's, with a larger manufacturing base and higher growth rates, regional military superiority (as we will see), and political motivations either way favoring military engagements. Taken together, these factors make for an unhealthy and potentially explosive mix.

America's Decline, Potential or Actual?

The relative economic decline of the United States has been a recurring theme in the public discourse for half a century. So far, such predictions have never become reality.

During the 1970s, the economy of the Soviet Union had reached a nominal size equivalent to over 70% of the U.S. Then it fell far behind and eventually collapsed. During the 1980s, Japan's economy had reached a nominal size equivalent to over 80% of the American. Then it started decades of near-stagnation and today's number stands at barely 25%.

And as of 2020, China's economy has nominally reached about 80% the size of the American. Does recent history repeat itself? Will there be another turning point that continues America's streak?

It certainly seems possible. China's many structural problems make it difficult to assess where its success story is heading. The times of the Covid-19 pandemic are untypical, but it still is notable that the growth gap for 2021 and 2022 has shrunk significantly.[72] Instead of the approximately 10% and then about 5%, we may approach a time when China "only" grows barely 2% faster than the U.S. - or not at all.

China's biggest risks come from demographics.[73] A shrinking and rapidly aging population can put an immense burden on a pension system that banks on investments into growth industries. Slowing economic growth rates aggravate this situation, particularly while large numbers of the Chinese population still are relatively impoverished. The country can probably speed up growth, but that may require a loosening of control over the economic sector. This relative laissez-faire approach had been the original trigger for China's growth in the 1980s and 1990s - but is the exact opposite of the current trend.

There are also reasons to question a repeat of history that compared the U.S. economy to the Soviet Union and Japan. Several recent actions of the United States have moved its economy toward more centralization, with the state, bureaucracies, and politicians taking on an ever-bigger role. Industrial policy and deficit-spending have combined with an intentional shift toward restructuring the economy to suit political objectives. Shifting power to politicians that pick winners in the economy often benefits large corporations or special interest groups instead of the most efficient organizations. America keeps deliberately expanding the world's most expansive and expensive (but not most effective) social system and the by far most expensive healthcare and education systems.[74]

This saddles the U.S. economy with huge bureaucratic overhead, often without delivering commensurate outcomes. As a result, the United States moves towards economic policies that seem more like those of China. For example, and right or wrong, it saddles the economy and society with rules that promote political priorities developed by elites, pursuing specific environmental, racial, and gender policies. This formally ideological mechanism potentially maneuvers America closer to China's political system dominated by what in the PRC's case is a communist elite. Such a development, however, would then likely also mimic China's challenges.

Somewhat ironically, though, it was the unleashing of non-ideological raw *capitalist* market forces that has worked so exceptionally well for China. By rewarding risk-takers, innovators, and cut-throat competitors, it achieved spectacular economic success and the widespread social advancement of almost all its previously abysmally poor. In the mid-1970s, China was one of the poorest countries in the world, with a mostly illiterate and often undernourished population. Since then, a political iron fist coexisted with an encouragement of free-wheeling capitalism. The PRC's leader of the 1980s and early 1990s, Deng Xiaoping, stated it as "to get rich is glorious." It hardly gets more capitalist than that.[75]

Still, particularly during the past few years, the Chinese Communist Party (CCP) under Xi seems to be bent on reigning in the economy and bringing it under political control. History suggests that such centralization of power and subsequent micro-management of the economy only works in theory, but not in reality. We would have to accept that small groups of politicians and narrow elites, and ultimately the leadership centralized in one person, can be more knowledgeable and capable than the combined self-organizing forces of society. There is little to no historical evidence that this can work, and definitely not over an extended period of time.

But the U.S.' recent practices follow in some similar ways. We continue pushing more money into outdated and vastly inefficient systems and institutions. They can in some ways be seen as comparable to China's state-owned enterprises. For example, we keep celebrating when more people get access to 19th century and early 20th century-style health insurances. These only deliver "sick-care," though, and recently have not even seemed able to increase life expectancy.[76] They often impede innovation because their interests are not outcome-oriented, meaning neither insurances nor hospitals or doctors get rewarded for keeping people healthy. Instead, they extract ever larger rents from society, keep bureaucracies happy, people employed, and generate profits for large and frequently quasi-monopolist private enterprises. Far from revolutionizing our systems and shifting from sick-care to future-oriented preventative personalized healthcare, the systems subsequently get ever more encrusted and unable to improve health outcomes meaningfully.

Lobbyism replaces competition and centralizes powers in the hands of ever fewer companies that do not add substantive value beyond the processing of paper and data. In healthcare, "payer" reimbursements employ millions and cost multiple percentages of the country's GDP, although this process could, in theory, be *completely automated.*

Another example is energy. Cheap and massive energy production and distribution contributed critically to the economic growth and rise of American living standards (and later also Chinese). As of the early 2020s, though, the U.S. government aggressively pursues the in principle admirable goal of a clean-energy transition in such a heavy-handed manner that it results in high inflation, voluntary dependency on foreign supplies, and a stifling of economic activity.

Technology may move the needle. We can re-invent our systems, from the production and distribution of energy to our supply chains, healthcare, and our communications and transportation infrastructure. The big question then is, which country and society, and which system, is better suited to succeed in this endeavor? And then, what does this mean for the geopolitical situation and the question of war and peace between the great powers?

China's recent pathway seems to give it a leg up. Only when digging deeper, the picture looks different.

Capitalism and Socialism - Here and There

Sometimes cynics ask why American capitalism has been funding Chinese communism during the past decades.[77] We kept spending money on Chinese products while running huge deficits, transferring hundreds of billions of dollars to Chinese creditors.

The simple part of the answer is that it made sense to both parties. It delivered more sales for China and lower prices for American consumers. But the more complicated part is that America did not do such a thing at all and that Chinese *capitalism* had at least as much to do with achieving rapid growth as the Chinese Communist Party's centralized controls and directives.

Much of China's export-led growth was based on a system in which the Communist Party and the government mostly got out of the way.

At a first glance, this may sound preposterous. Obviously, the Chinese government controlled the economy and aided its businesses at large scale. The Communist Party kept its power and played an important role behind the scenes. Politicians, including many in the billionaires-club that the Politburo essentially had become,[78] skimmed off a sizable junk of the profits. Their actions often resembled those of venture capital firms that open doors and ensure access to capital for receiving disproportional rewards. In China's case, these "venture communists" just had less risk because they also could impact the regulatory environment and laws affecting their interests.[79]

The Chinese government also demanded corporate structures that favored Chinese owners, and regulations that always supported or even outright demanded knowledge transfers. Organizations closely associated with Beijing actively participated in espionage, and protected copyright infringements and price dumping, often on a massive scale. This supported Chinese strategies to copy and underbid whole industries that eventually all but disappeared from the West.

A big part of the PRC's approach was like the one previously pursued by the four Asian Tigers, South Korea, Taiwan, Hong Kong, and Singapore. The government built roads, airports, and rail systems, successfully negotiated China's joining the World Trade Organization (WTO), and kept its massive state-owned enterprises. In high-growth technology industries, though, it mostly got out of the way. As long as it did not challenge the power of the Chinese Communist Party, the PRC accepted the premise that an open economic system, aided by private and government-supported corporate espionage and legal protection, was the answer to feed even such a populous country, and to educate and enrich its people.

Following a role traditionally also taken on by Western governments, for example, China used its swelling coffers to build the World's most advanced infrastructure, faster and on a larger scale than any country ever did throughout its history. But, in general, the government did not direct specific corporate actions.

Although the Chinese government actively supported startups and ventures, these then had to compete aggressively among each other. Only the winners gained access to a vast and rapidly expanding market. Such competition was most noticeable in the high-tech sector, where lots of entrepreneurs copied and then adjusted and enhanced successful Western business models.

Some with first-hand knowledge saw the business climate in the most innovative and most rapidly growing parts of the economy resembling capitalism more unfettered and raw than anything that has existed in the U.S. during the past half century, with the potential exception of parts of Silicon Valley. Others go further.

For example, AI and China expert Kai-Fu Lee, a venture capitalist playing a central role in the emergence of AI both in Silicon Valley and in China, describes the difference between the U.S. and China as the difference between cooperative competition - and *war*.[80]

If Lee is correct, China's rise would therefore be mostly a story of capitalism's power and ability to successfully transform societies. And people in Latin America and Africa seeking rapid advances of prosperity would not hesitate to welcome the results. There is probably no country anywhere in the world that would not gladly accept most of the outcomes, like average annual growth rates of above 10% for decades, a doubling of GDP approximately every 7 to 10 years, and an overall 50 to 80-fold increase of GDP over 40 years.

Even more so than the rise of the United States and its creation of the first large-scale middle class in human history, China's economic ascent has been a dramatic success story that at least temporarily benefited most parts of the world. Mining and energy suppliers sold unprecedented amounts of raw materials while consumers got lower-cost products because of global supply chains.

Unlike the U.S., much of China's meteoric rise to economic superpower status was export-driven. An almost unlimited pool of low-cost labor made it economical to use China as a basis for most mass manufacturing. After the turn of the century, clusters of manufacturing capability turned China into the default location for most new manufacturing in the world.

The huge inflow of hard foreign currency, mainly U.S. Dollars, generated a dramatic rise of the still very low incomes of the Chinese people. It facilitated the education of the masses, while also enabling the country to execute the most massive infrastructure build-up of any society ever in history.

The relationship between the U.S. and China turned into a uniquely symbiotic one. Notably, the PRC generated the overwhelming portion of its huge export surplus with the U.S. and only minor ones with other countries.

Not only were China's growth rates phenomenal, for decades mostly above 10%, but its economy became ever more mature and increasingly domestically oriented.

With about $2.5 trillion in 2019, Chinese exports accounted for 17.5% of the country's economy.[81] This is very significant, but the domestic market's share keeps rising. Just 10 years earlier, exports accounted for almost 24% of the economy. Today's numbers are getting close to those of the U.S., whose exports are slightly below 10% of its GDP, with imports being above 12%.

But even more importantly, the PRC's trade shapes supply chains of physical goods, including energy, raw materials, and semi-finished and finished products. And that creates *mutual* dependencies.

Hard and Soft Powers

If the primary goal of economic activities is to create hard and measurable wealth and power, the West frequently indulges in what we can describe as secondary priorities. Above all, the regulatory regimes in the West are exorbitant. Often, their goal is not to eliminate poverty but to create equality, not to maximize social good but to engineer it in a specific form, and not to maximize health and education but to do so in a way that is left to often inefficient and highly regulated "markets" of special interests driven by election dynamics. Even the most well-meaning programs and the corresponding laws, taxes, and regulations can complicate and slow down growth.

Expressed somewhat simplistically, China's approach is to keep the communist party in power but otherwise subjugate near everything to economic growth. It lays the foundation to raise all the boats, lift hundreds of millions out of poverty, bring them into the middle class, and then let them gain genuine wealth. The condition for this is the use of vast amounts of energy, the destruction of the old, and the building of the new.

This is the positive view. There is also a negative one. As just mentioned, a few pages ago, most of the 153-member Chinese "parliament" are billionaires.[82] They are about as many among this tiny political elite as all those living in the United Kingdom and Japan *combined*.[83]

Political freedom and free speech are limited, and any kind of dissent can trigger draconian punishment. The overall results of China's economic model, though, are exactly what most of the rest of the world wants to have, particularly in low-income countries. And if this comes with less political freedom, many seem okay with it.

Capitalism laid the foundation for a doubling of GDP approximately every seven to ten years, for about four decades. Obviously, it worked for China. Businesses and entrepreneurs generated tremendous wealth. Even in one of the most unequal societies of the world, the trickle-down effects of an unequaled economic boom pulled between half a billion to one billion people out of poverty, often out of *abject* poverty.

This also transformed the country into a major global power, starting slightly before the turn of the century. First subtly and then ever more openly, for about a quarter century, the PRC then used part of its new manufacturing capabilities and economic power to continuously strengthen its overall geopolitical position. And subsequently also its military might.

It did not go together with the political opening and establishment of Western-style democracy that many in other countries had hoped for. But the overall economic success is undeniable. Only recently, the growth rates have shrunk to about 5% to 7%. It is still unclear whether this is *despite* or *because* of an ever more assertive role of the CCP and of President Xi in economic and political matters, or for other reasons altogether.

Toward the Future: Growth Scenarios

The primary determining factor of military power is economic and manufacturing capability. This is driven by innovation and technology.

Measuring economic capabilities and power, though, is complicated. For the purposes of the rest of this chapter, we will stick to the crude measurement of GDP. As insufficient as it is, in many ways, it still gives us an excellent idea about economic power, particularly *relative* to other countries. The main comparison here will, of course, be between China and the United States.

In the decade before Covid-19, the U.S. economy grew on average by 2.3% per year. In the same time span, China grew at an annual rate of 7.7%. 2019 was the only time during the past 30 years that China's growth rate dipped below 6%, and only barely. We can now run three scenarios.

Aggressive Growth Assumption

If China would grow at a rate of between 5% and 6% annually while the U.S. remained at or below 2.4%, within 15 years China's economy would be larger than the U.S. even in nominal terms. If this were to continue until 2050, the U.S. would have doubled its economy compared to 2021, while China's would be between 4.1 times and 5.4 times as big as it is today (depending on whether assuming a growth rate of 5% or 6%). In PPP terms, China already has a larger economy than the U.S. By about 2040, China's economy would even nominally be *twice* as big as the U.S., or about as big as the U.S. and the EU *together*. This is less than 20 years from now! By around 2050, it would be as big as the U.S., the EU, and Japan together.

Nobody alive has experienced a time when the United States was not the biggest economy on the planet. The scale of the difference may therefore sound absurd. But it is not.

Even if we extrapolate the numbers all the way to the year 2050, as I just did, the average Chinese citizen would still be much less wealthy than the typical American. Although China's economy would be about twice the size or even more of the U.S., it would take three more decades of similarly diverging growth rates to catch up to the per capita income level of an average American today. In the meantime, though, Americans would have become twice as wealthy.

I labeled this scenario "aggressive," although that is relative. It already assumes about a *halving* of China's average GDP growth rate of the past 30 years! Despite this significant reduction, and even while its citizens remain far less wealthy than America's, China's overall economic power would in 2050 still, by far, eclipse the U.S. A scenario that turns China into an economy twice the size of the U.S. by 2040 is therefore far from implausible. It also seems almost exactly what not just Chinese leaders expect to happen. The Chinese public and most people in East and Southeast Asia, and some other parts of the world, see exactly this scenario.[84]

Moderate Growth Assumptions

Staying much more conservative, a Japanese study predicts that by 2060, China will be about 20% larger than the U.S.[85] Another 2020 study of the Japan Center for Economic Research predicts China getting close and surpassing the U.S. but then falling behind again by the middle of the century.[86] Demographics would be a major driver for this switch back to the U.S. leading. My back-of-the envelope calculations of such a scenario come to a similar result. They assume that the old 10% growth rates will remain history, and that the numbers will be cut to half. This is close to today's rates, although slightly lower.

But Chinese numbers are, for political reasons, likely artificially inflated. Since this has possibly happened for a while, we also must shave off a portion from today's official GDP numbers. Economics professor and Bloomberg View contributor Christopher Balding has been writing about this for years.[87]

A 2019 paper by Professor Chang-Tai Hsieh of the University of Chicago, Zheng (Michael) Song of Tsinghua University and the University of Hong Kong, and University of Hong Kong Ph.D. students Wei Chen and Xilu Chen concludes that China's annual growth has been artificially inflated by about 1.7% for many years.[88] The total GDP number may therefore be around 20% lower than officially reported. If we assume this, China will still continue to close the gap. By the middle of the century, it would reach about parity with the U.S.

According to this scenario, the average American would be, in 2050, about four times as wealthy as a Chinese citizen. Both nations' overall economic power in nominal terms, however, would be approximately equal.

Pessimistic Assumption

We can also assume a more pessimistic development, from China's perspective. What if the so-called "middle income trap" springs, a rapidly aging population overburdens the social system, and China's real estate, infrastructure, pension, and finance bubbles burst? The economy may enter a phase of relative stagnation, requiring painful adjustments across all sectors. In this scenario, political unrest spreads despite increased surveillance and directed social modeling, and the effectiveness of technocratic measures dissipates.

All this may happen as the working population, and then the population contracts. China's growth rates may then fall precipitously and hover at somewhere between 1% and 2% for the better part of the next three decades. These numbers may be lower than those of the West, or at least the United States. With a falling and aging population this would still mean a consistent increase of per capita income and a continuing rise in economic productivity.

Such a scenario is not impossible either. Many in the West, including myself, continue pointing at the structural weaknesses, challenges, and outright threats to the continued success of the Chinese model. The power trajectory may quickly change, by any of several bubbles blowing up. Among the likely culprits are infrastructure investments without viable business models, pension funds in the middle of a retirement boom, housing and real estate markets, local and corporate debt, oversubscribed financial markets, and general inefficiencies arising from too much centralized planning and political meddling into markets.

For the Chinese, all this would likely be a disaster. They just would have lifted about a billion people out of poverty, build the largest middle class in human history, and all that faster than any other large society before. But now most of its people would hit a brick wall and could no longer noticeably advance.

For the geopolitical position of the United States in relation to China, it would be a boon. We would see, for the third time, after the Soviet Union and Japan, a nation outgrowing the U.S. and reaching about 60% to 80% of its nominal economic power, just to then fall far behind again.

However, this scenario would require a complete reversal of a four-decades-long trend. It would also mean that an almost all-encompassing literacy of the population, a strong focus on education, the training of an advanced workforce, the largest manufacturing base in the world, a diverse economy, overall outstanding infrastructure, and significant advances in technology would not be sufficient to maintain a decent growth rate.

And therefore, a purely economic or social "middle income trap" seems unlikely to me. Although something similar happened in Japan, China's domestic market alone has a ways to go. This makes its society more open to radical change using sweeping technocratic methods. And we must not discount the use and effectiveness of technology either.

In principle, I trust the efficiencies created by distributed and diverse self-organizing systems of democracy and capitalist competition. Rather than some ominous "middle-income-trap," I see the biggest threat to China's model and success more likely in its government's abolition of market dynamics. Recent decisions suggest that the PRC now thinks it can improve on market's efficiencies and instead direct future success top-down.

Overcoming New Self-Inflicted Challenges

Since the mid to late 2010s, heavy-handed authoritative and totalitarian measures have affected Chinese businesses and even started to deliberately restrain some of the most innovative, successful, and admired business leaders, like Alibaba's former CEO Jack Ma.[89] This is not just a political concern. More party control and authoritarianism will likely also have a negative economic impact and reduce domestic innovation. Because matters of the economy are too numerous and complex, top-down approaches cannot work as efficiently as when systems self-organize, for example, through markets.

Instead, centralization usually results in a misallocation of resources, particularly if the "guidance" (or "interference") comes from a political class and not an economic one.[90] China's increased political authoritarianism also triggers pushback from other countries. It generated a growing awareness in the West and even among many Western business leaders that political and philosophical differences matter.

Even today, though, and except for the erratic fluctuations during the Covid pandemic, Chinese economic growth rates still exceed those in the West significantly. In principle, this continues to be good news.

During the past decade, a much wealthier and more highly educated Chinese population approached the status of a global middle class. In combination with good infrastructure, this makes brainpower accessible. And in the 2020s, human brainpower counts more than ever. To continue growth requires establishing a system that taps into the creativity and innate drive of people. This is the key challenge for most country's future, and China has made some strides in that respect.

Geopolitical Consequences

Encouraging the above pessimistic scenario would be cynical, considering its consequences for the Chinese people. It is not the most likely one, anyway. The middle-income trap affected many

Latin American countries. Most of these, though, neither had at their avail China's huge markets, nor the level of world-class infrastructure, comprehensive education, broad manufacturing base, vast capital accumulation, economic diversity, and technological co-leadership role that the PRC enjoys today.

More likely than relative stagnation or regression, we will therefore see a mix between the above moderate and the aggressive relative growth scenarios. China will experience internal and external shocks that require it to make some painful adjustments. Growth in the U.S. will not all be smooth either, though, with the country struggling with its own challenges. We should assume that China's growth rates will probably stay slightly above America's, somewhere between 1% and 2% higher. If we also assume a moderately lower baseline due to politically overstated GDP numbers, it roughly suggests the following geopolitical situation.

In short, we can apply an approximate "rule of 1.5." According to this, by 2050, China's economy would be 1.5 times the size of America's. Compared to the "rich billion" of North America, Europe, Japan, South Korea, Australia, and a few smaller liberal market economies combined, this Western world's totals would be overall 1.5 times as large as China's, with a population at least twice as wealthy. These numbers would by then be similar for either PPP or nominal GDP.

The United States should consider this rough scenario as most probable because it clarifies its geopolitical options, particularly in East and Southeast Asia. Without being overly dramatic in either direction, it gets the U.S. mindset ready to seriously consider a changed world. If America pulls ahead again, no foul has been done. Otherwise, we are at least mentally and strategically prepared for how to deal with a China that may become Number One by some critical measures.

Either way, though, the most probable scenarios will continue to affect the military balance in East Asia. China is likely to continue translating its increased power into even more military might, especially close to its shores. A United States that keeps spreading its military commitments and economic activities around the globe, in the mold of today's alliances, global institutions, and military strategies and weapons, will find it next to impossible to keep a credible counterweight in China's neighborhood.

This begets the question of what can, or should, be done about all this.

In the short term, we have a few urgent priorities. The United States has large dependencies on supply chains involving China, from semiconductors, microchips, rare earths, medical supplies, and a myriad of semi-finished products. Western corporations still keep teaming up with China in deals that continue transferring technical and business know-how and intellectual property (IP) to China. The basic arrangement is for China to give access to its markets and lower-cost production facilities, in exchange for building manufacturing capabilities, training the Chinese workforce, and often even outright transferring IP to Chinese authorities or majority business partners.

There are some good arguments to cautiously continue with a more balanced version of some of these practices. After all, Silicon Valley did not only perfect competition but also cooperation even among competitors. And China's copying, and the protecting of businesses and industries, is not too dissimilar to how other countries have developed, including the U.S. Stunningly different are just the scale, brashness, and speed, and the government in Beijing deliberately supporting it. In the medium and long term, though, the current specific form of IP extraction, and its scale, is unsustainable. It keeps hollowing out Western capabilities.

Part of the reason is the special public-private partnership model that the PRC has near-perfected for decades. It included not just the government's support of spying through paramilitary groups, but also a regulatory framework that ensured the transfer of IP and the corporate laws to ensure financial and competitive advantages. And then it provided access to large markets and to low-costs manufacturing using state-owned enterprises. The government then played wingman with global propaganda initiatives and measures. And throughout that process, the Communist Party and its governmental executive leadership defined the priorities and general direction while making sure that their control and ownership remained absolute.

Of course, we in the West must look at what works and why. There also is nothing wrong with learning and even adopting certain components like extreme forms of competition to drive innovation. But, why would we want to become more like China to fight them? We would have to believe that they have a better system. Do they?

And this is exactly the question that we must answer.

Shortening innovation cycles driven by exponentially growing digital technologies give most societies, almost constantly, a chance to start on the bottom of a new wave with large-scale growth potential. 5G, 6G, satellite constellations, AI, quantum computing, IoT, digital money, blockchains, robotics, and biotech. The tools are all there to build or rebuild economies and societies. For geopolitics, this means that there always is a new deal, a new opportunity to be successful, and therefore also a way for societies to rise and fall compared to others. This also applies to the U.S., which starts with many advantages to make use of modern innovations.

Our likely most promising course of action is, therefore, not to copy Chinese top-down strategies but to focus on our traditional strengths. It means adopting more competitive forces of the kind the country unleashed during both the U.S.' and China's most aggressive growth phases. We can balance this with new models, e.g., using tokens and blockchains, for governing common goods so that they benefit everybody in society. This would enable us to grow ourselves out of relative decline, ideally together with other liberal and market-oriented democracies.

I will address this question in much more detail at the end of this book. But the biggest driver of progress must be at the front and center of this process: technology and particularly the digitization of our world.

Before we get there, though, let us explore how soft past and current transitions have impacted the world's perceptions of America and China.

6

Soft Power

Nobody Loves Hate, or: The United States of China

The free world changed in 2001, in two steps that were not even 50 days apart from each other.

First, there were the 9-11 attacks, when 19 extremists destroyed the World Trade Center in New York City and attacked the Pentagon. Almost 3,000 people died, and the economic damage was huge.

But something much more consequential for America's power followed 45 days later. On that day, a new law became effective with the support of the overwhelming majority of both parties in both houses of the U.S. Congress. It was the Uniting and Strengthening America by Providing Appropriate Tools Required to Intercept and Obstruct Terrorism (USA PATRIOT) Act ("Patriot Act").

Together with the way the U.S. conducted what it soon labeled the "War on Terror," the heavy-handed interpretation of the Patriot

Act deeply darkened the view of America in many parts of the world.

Of course, defending America was justified. But these decisions, and an ever more acerbic and even hateful domestic political discourse shared with the rest of the world, had profound consequences. They put the U.S. on a path to lose the "soft power" advantage it had for such a long time. Joseph Nye, former Assistant Secretary of Defense and Dean of the John F. Kennedy School of Public Service at Harvard University, had coined this term and also chose it as the title of a 2005 book.[91] It described how the universal appeal of American values and principles could compensate for or translate into the hard power to make other nations do what otherwise would not come naturally to them. Mostly, this would cause other countries to pursue similar policies and be, in principle, more understanding and supportive of American foreign policy actions and strategies.

American soft power is still immense. But China has made huge strides in that department as well. The success of the "Beijing Consensus" can increasingly compete with the "Washington Consensus."[92] A large reason is that China's rise has affected many people where it really counts, often physically and in their economic well-being. And another one is that America's conduct abroad and the way it portraits its own country are simply unappealing for many foreigners.

Today's Disconnect: Ukraine and the World

Russia's brutal and unprovoked attack on the independent nation of Ukraine showed the result. A common narrative in the Western media is that the world united and acted together as strongly as hardly ever before.

The world, really? As so often, the truth is more complex.

Russia's war united NATO and the core Western nations. The EU, Japan, South Korea, Canada, and Australia, together with many other nations, particularly in Europe, levied harsh economic and political sanctions. Some of these extended to Western-dominated international organizations. The public in many European nations, most notably in Germany but also in Finland and Sweden and even Switzerland, was shocked. Businesses cut ties with Russia or even directly supported Ukraine. Western governments introduced significant changes to their defense policies and openly supported the Ukraine with the delivery of weapons. Long-time neutral Finland and Sweden even applied for NATO membership, fearing to also be in Russia's crosshairs.

Indeed, "world opinion" and the "international community" stood solidly against Russia and for the rule of law and democratic freedoms. As long as one defines "the world" as "the West."

The rest was far from united. If "world opinion" means "the opinion of the majority of the people living in the world" then it may have been a draw, if even that. For one, China not only stayed on the sidelines but even openly supported Russia economically and politically. India, Brazil, and Mexico refused to condemn Russia. The Pope condemned not Russia's unprovoked brutal crossing of a border with thousands of tanks and tens of thousands of troops, and the bombing of infrastructure and civilian populations. Rather, he blamed Western political actions. Saudi-Arabia and the United Arab Emirates (UAE) refused to take phone calls from the President of the United States. Even Israel tried to mediate, as if anything about Russian actions could seriously be subject to interpretation. Pakistan, Bangladesh, and Indonesia were other large nations abstaining.[93]

Africa stayed mostly quiet, not the least for one important reason. The most brutal and deadly military conflict of 2022 is not taking place in the Ukraine. Rather, it is in Ethiopia, with multiple times Ukraine's death rates, but widely ignored by Western powers and people.[94]

World opinion was, therefore, in reality the opinion of wealthy countries with a still overwhelming share of economic and military power, feeling a cultural affinity for a European nation. But if this is indeed the measuring stick, then China's recent advances and the West's perceived relative decline are disconcerting.

On the surface, this seems surprising, but when we look deeper, we better understand at least some of it.

Why America (and the West) lost their appeal

News from America is the most widely watched and communicated foreign news anywhere in the world. However, particularly during the past few years, widely propagated by America itself, extremely unappealing accusations about the United States dominate the headlines. It has been full of self-incriminating themes and reports. These purport widespread racism, inequality, homophobia, antiforeign resentments, and a discriminating society in which free-wheeling capitalism enriches the few while impoverishing the poor.

As a widely traveled immigrant who has worked in many countries, done business in dozens of them, and who has lived on three continents, though, this is not what I see. All the above exists, definitely, and we must continue to improve on it. But it is not *typical* of American society, and not more so than in almost any other society on Earth. Neither my personal experience nor the data I see support sweeping claims of aggressive and systemic discrimination and human rights violations.

However, when the Biden administration had its first top-level meeting with their Chinese counterparts, it sounded like that. They had to listen to a long list of accusations about exactly such extensive human rights violations across the United States.[95] America's enemies, and China, do not even make up such claims - they simply copy them from the extremist partisan claims, or single-issue political movements that dominate the political debate inside the U.S.

This has a direct impact on our foreign policy and on our global appeal. When part of the American political spectrum accuses their country of despicable acts and systemic actions with words equivalent to those used in the past when describing oppressive totalitarian regimes, then we have a serious problem for our foreign policy. And if both major sides in American politics portray each other with disgust and utmost negativity, this contributes even further.

For the outside world, it is Uighurs here vs. Racism there. Brainwashing and social credit systems here and wokism and environmental, social, and governance (ESG) criteria or climate change activism there. Inequality in China here, inequality in the U.S. there.

Many living in the poor nations of the world want what America claims to be about. But many in the U.S. say their own country cannot deliver. China shows it can, at least for most of its massive population.

At best, we have a wash here. In the eyes of many who visit or work in China, they may even consider it being ahead of the U.S.

Uighurs and Western Versions of Brainwashing

Many call it genocide what is happening in Xinjiang. I do not like the use of this term because it tends to belittle the word's traditional definition of killing the overwhelming majority of an ethnic group, like during the holocaust. And this is not going on in China. But it is bad, very bad, to say the least. I definitely see a horrible misuse of governmental power and a massive violation of basic human rights. How else can one describe the PRC's oppressive authoritarian measures, including torture, forced labor camps, and mass-executions, and its aggressive re-education and brainwashing tactics in Xinjiang?

It then does not help, though, if we lose perspective and exaggerate aspects that supposedly are unique to China (and North Korea), but in reality aren't. In "The Perfect Police State," Geoffrey Cain describes surveillance methods as "barbaric" that include facial recognition in gas stations or privacy-invading apps tracing online activities, social encounters, and physical movements. As a decade-long student of that topic, and a member of the International Association of Privacy Professionals (IAPP), I am familiar with the technologies and systems Western businesses use, and particularly giant gatekeepers like Facebook or Google. Compared to them, descriptions of video cameras in Chinese businesses and of their consequences sound "cute" contrasted to what is already commonplace in our own society as well.[96]

As a matter of corporate policy their business models depend on what in any other context would be called totalitarian surveillance and manipulation. Google and Meta (owner of Facebook, Instagram, and WhatsApp), and others including Microsoft, Amazon, and Apple, created profiles of every person or device that ever came in touch with any of their products. These are billions of people, including nearly everybody in the West. Then, they collect millions and billions of datapoints correlated across devices and covering all aspects of each person's lives. This is not just data generated in their apps, far from it. With the partial exception of Apple and Amazon, they also pull unrelated and personal or otherwise sensitive data from computers, phones, and other devices. Think emails, browsing actions, mouse movements, keyboard clicks, location data, words mentioned around intelligent assistants, and so much more.

And then, Google and Meta make this data available to whoever pays for it (or otherwise compensates them for it, which will be my next point), to influence their actions and thoughts based on prediction algorithms. Nothing I just said is speculation, accusation, or even just interpretation of actions. I am simply reciting publicly stated and known facts, stated by these companies themselves. They may often be obscured, and their interpretation may be spun. But the facts themselves, those I just described, they are undisputed.

For this they get legal and political cover from law makers and governments. These not only look the other way. It is illegal for governments to do what Big Data does - but currently not for businesses. And so, federal agencies strike deals with these surveillance capitalists, like their counterparts in other countries. And there we go, all is legal. Wink, wink, nod, nod. No harm done, right? Welcome to the United States of China.

And also in the West, political correctness as reflected in ESG policies and corresponding or related political activism try to top-down re-engineer society based on concepts developed by small groups of elites. Much of this can be difficult to differentiate *in principle* from what the Chinese Communist Party does. Both intend to control a certain flow of information and opinions, stifle opposition, and bedevil those arguing for alternatives. And then they try to use governmental power to enact legislation that aims at cutting off investments in economic activities that until just recently were the very foundation of Western economic and social successes. These same methods are also used by single-issue extremists of any direction, motivated about topics ranging from immigration, abortion, to free speech.

In the meantime, I have no reason to believe that the politburo in China get up in the morning and say, "how can we torture as many people as possible" or "how can we put more people into concentration camps and create a horrible life for them." Rather, I find it much more plausible that they honestly believe their policies maximize, in a utilitarian sense, the benefits of the greatest numbers of Chinese people.

And, of course, most of these leaders also want to continue gaining and enjoying vast powers and riches for themselves and their families. By that measure, though, the result can seem on the surface, again, similar to elites in many other countries. Elites and rich people are here and there. The casual observers do not realize, or don't care, that only one is based on force and its people cannot change their rulers. What they see is that Western "elites" also believe that they have "all" the answers and are better suited to deliver good, or even the best possible, results for society. Despite their claim to live in a "democracy," many of these Western "elites" also seem to honestly believe that they know better than the ignorant masses. They then also think that they deserve their powers and riches, and that their policies should be implemented even if most of the population does not agree.

In reality, and in the end, it comes down to a matter of degree and to the processes and their balances. And these counterbalances don't exist in China to the same degree as in the West. Some exist, but they are much easier to overcome. The West, on the other side, has built-in change of government and distributed power, even though with limitations. These all-important nuances get, all too often and easily, lost when foreign news report on U.S. politics.

And so, of course, I do not intend to equivocate. A liberal world is not just more favorable for humans, at least in the long term, but also likely in most situations. It also is a superior way of organizing social systems and delivering economic results. As outlined in chapter 5, the economic system can make a huge difference. And in chapters 14 and 15, I will describe why and how a re-organization of our society can continue to deliver on this.

But with all this being said, we must watch out and stop aggregating power in small economic and political groups of society as we have been doing for decades. Continuing this process would contradict the very philosophy that underlies our own society.

There are other indicators for where we risk going wrong. Self-anointed unelected administrators accrue political and regulatory power in bureaucracies. They set rules and in essence make laws even when neither being elected nor being part of the law-making branch of government. And even inside that branch, continuously re-elected politicians accumulate political power and gain positions with disproportionate levels of influence compared to their formal mandates. In many or most cases, neither of them has substantive hands-on experience in the matters they legislate and regulate. They are administrators and lawyers, fund-raising machines, not subject-matter experts or entrepreneurs operating in highly competitive and rapidly changing environments.

"It's good to be king!"

Mel Brooks kept repeating this statement in his comedy "History of the World, Part I."[97] And America knows the feeling as well.

During America's unipolar moment, when the Soviet Union collapsed and Francis Fukuyama talked about the "end of history,"[98] everything just seemed to make sense inside the Beltway of Washington, D.C., and far beyond. The free-market economic system had showed its superiority. The rules-based international liberal world order had won. Democracy and individual freedom dominated the planet. The United States of America, Washington, was on top of the world, with unrivaled economic and military power.

Humility no longer was part of the political vocabulary. Instead, we got the "Washington Consensus."[99] This set of principles and strategies focused on a rules-based free-market economy roughly modeled on policies and institutions like those in the United States. But humility may be exactly what we need, particularly today, to distinguish hubris from lasting greatness.

For this, it is helpful reminding us of another Washington, not the city but the person they named it after.

"If Washington does that, he will be the greatest man in the world," said the British monarch.[100] In 1783, George Washington had stepped aside from a position of supreme power. Out of his own volition, he became a private citizen again and let the democratic process in a new nation find its way.

200 years later, to America's credit, it did not really pursue empire either, even after the Cold War was over. Rather, the U.S. kept its alliances and military capabilities, and used them for idealistic purposes when it deemed so necessary. But it also became self-absorbed, projecting its internal struggles and issues into the wider world. Part of this led to the continuing pursuit of shaping the world to mirror American or Western values and solutions.

Aren't We the Good Guys?

Of course, like most powers since ancient times, we say we must aggressively protect our dominating position because we are morally superior and "the good guys," with the future of civilization depending on our continued success and hegemony. This may even be correct, at least when referring to our values and formal priorities ("life, liberty, and pursuit of happiness"). But it is not necessarily how many in the rest of the world see it. A lot of this is because of the effects of China's economic growth while the U.S.

kept pushing its ideological priorities when dealing with other countries.

China's ascent has made the rich world richer, first by cutting manufacturing costs and then by letting outsiders sell products into a mushrooming Chinese market. For the poor, though, the impact was often even more substantial.

A briskly rising China lifted all the boats, rapidly. Since before the turn of the century, China has been a primary engine of global growth. Soon after, it became the biggest one. At home, this raised almost one billion people out of abject poverty and elevated most of them into the middle-class. It was "trickle-down economics" on steroids.[101]

Abroad, this became a boon for countries with lots of natural resources, like those in Latin America, Africa, and the Middle East, and for China's other trading partners. They benefited from exporting natural resources and importing lower-cost products, including mobile phones and technical equipment. When in 2006 the so-called BRICS countries gathered (Brazil, Russia, India, China, and South Africa), most of their power and dynamics came in reality from Chinese above-10% annual growth rates. Indian economic liberalization aided as well. But the other three, Brazil, Russia, and South Africa, were mostly suppliers of raw materials and energy, primarily to China.

Some of China's neighboring states feel threatened by China. However, they are also in other ways strengthened. China is East and Southeast Asia's main trading partner. It is a key part of the supply chains for some of their most important products, and a principal source of foreign investment. This even applies to antagonistic Japan and Taiwan. And chances are that many among their population fear that stopping China politically and militarily will threaten their own chances for prosperity as well.

The PRC has demonstrated a rapid way out of poverty, and onward to wealth for their people. And other Asian nations, like many in Africa and Latin America, realized that since the turn of the century, they did not have to strictly follow the so-called Washington Consensus mentioned earlier.

Somewhat ironically, China's rise proved the power and universal appeal of capitalism. On a personal level, it is clear for most of us that economic success is a key condition for really being free. If you are genuinely poor, you are not in control of your circumstances. The PRC delivered big time on this. It got more people out of poverty, and more rapidly, than any other society in human history.

At least to a degree, capitalism has the same effect on entire societies and nations. Economic capability and might precede military power. It is a necessary condition. This is why preventing another society's economic development would also limit its military power.

Therefore, in the eyes of the poor of the world, including those living in China's shadow, it makes them nervous when they hear about a strategy of containment of a China that has not done them any obvious harm but delivered many benefits that they deeply feel in their daily lives. When they see China being challenged, "contained," or "denied," they do not instinctively cheer. Such strategies do not win hearts and minds in Southeast Asia or in Africa. People there wonder what it means for them, personally, and for the chances of their societies to develop.

The U.S. must not discount how the rest of the world senses this. We are walking a very fine line here. Points about being the leader of a free and rules-based world and advocating for individual and economic freedom do not sound as appealing as they used to. Their backdrop is U.S. troops being spread around the globe, fighting wars in foreign countries, the U.S. government and American businesses surveilling all electronic communication around the world, and U.S. military, tax, and regulatory authorities intervening in domestic matters of sovereign nations.

Economic Competency

All this may still not matter much were it not for a growing perception that the U.S. and the countries of the liberal market-oriented Western world would no longer be able to effectively deliver prosperity for their people.[102] Willpower and mindset may be the underlying factors to determine whether we succeed with our economic aspirations. But tangible value and wealth are the most visible and relevant measuring sticks for whether it worked. And although Western prosperity keeps growing, albeit slowly, it increasingly seems ineffective in confronting the most crucial challenges of the world.

Much of this may be because of the eternal U.S. election campaigns that swamp the world with negative news about even the most venerable American institutions. These often grossly exaggerate relatively minor issues and challenges. But this is not all. Some numbers support some of the criticism claiming an American lack of economic competency, and not just on the macro level. There, governmental taxes and collections go up, the regulatory burdens for businesses expand, and the U.S. national debt increased to the highest levels since World War II.

But even before Covid-19, Americans experienced issues and limitations in their daily lives that other advanced societies did not seem to have to the same degree. Life expectancy has been stagnant for about a decade. The U.S. barely makes the top 40 in the world anymore.[103] Despite the country arguably having the best hospitals and a globally unsurpassed urgent care system, the health of average Americans seems deteriorating. Rates of chronic diseases, diabetes, obesity, and mental illnesses keep rising.

In many places, across the country, infrastructure seems crumbling. Roads have potholes, bridges teeter on the edge of collapse, and airports remain outdated. Repairs, renewals, and expansions seem to take forever. Maybe the 2022 infrastructure bill will fix this. However, only a portion of the approved funds is to pay for tangible physical output.

The outcome in the education system is more mixed. American universities still top global rankings. Up to high school, though, the picture is not as positive. Some results of international standardized tests are also concerning and widely communicated across the world.[104]

There are encouraging counterpoints for all of this. But the real whopper is the associated non-value adding bureaucracy, and the out-of-pocket costs incurred by American taxpayers. Often, the government and politicians work hand in hand with business and labor interests, like in healthcare, education, the energy sector, transportation, and recently also the large tech companies.

The result is discouraging. Healthcare in the U.S. costs almost twice as much as in the next most expensive large country in the world. Prices still keep rising. Colleges can raise prices disproportionally compared to other industries. Air travel and Internet access are less competitive and more expensive than in Europe. [105] And an often restrictive and confusing regulatory burden complicates and stifles competition and innovation.

U.S. President Eisenhower warned of a military-industrial complex twisting defense priorities and procurement decisions in their favor. Nowadays, though, there also seem to be similar "complexes" in healthcare, pharmaceuticals, education, finance, energy, and green industries. In some other countries, similar practices and maneuvers that benefit special interest groups are called "corruption." In the U.S., the preferred word is "lobbyism."

This may be an unfairly harsh characterization of the American system. Indeed, it leaves out many positive indicators and developments. But with soft power, it matters what the rest of the world *thinks*. Foreigners notice that in many important fields, the U.S. still often seems ahead of other large countries. Then, though, they wonder why many Chinese airports, roads, and railroads are more modern, living expenses are lower, life expectancy is rising, and healthcare and drugs are cheaper.

Not just the United States' military strategies and self-image seem stuck in the 1950s. The question of efficiency and sophistication comes to a head when visitors notice Americans paying with personal checks (!), stuck in the mid-20th century. We see practices that most other societies have passed decades ago or skipped altogether. In China, as in some African countries, almost everybody uses their mobile phones to pay.

The Destructive Effect of Never-Ending Political Campaigns

And then, social media-driven political polarization spills abroad and influences public opinion outside the United States. Foreigners quickly take sides in America's domestic political debates, even when they only partly understand them. They are interested because America often seems to affect them personally. Of course, U.S. politics rarely is as simplistic as foreign media describe. But, as one can easily see when traveling outside the United States, the effects often are devastating for America's overall image.

People abroad take their clues from billion-dollar campaigns dominated above all by destructive, and deliberatively selective and emotional opposition research, negative advertising, and the bedeviling of the "others." They see so-called "independent" media take sides and big technology companies censoring opinions and even banning books in the name of "battling fake news." All parties keep questioning election results: in 2000, 2016, 2018, and 2020. They belittle the democratic and election process in a highly incendiary manner. And then they accuse each other of manipulation, corruption, and other illegal acts. Many who take such partisan propaganda at face value wonder how much this so-called democratic and free country differs from the ones it opposes.

Foreigners take notice when mass protests, violence, and chaos seem to reign in many U.S. cities. They see when police withdraws while, or after, being accused of systemic abuse of power and racism. Then they read about crime rates spiking and wonder what is going on. They see how relatively rare acts of injustice, or of extreme political positions or actions on the fringes, or the thoughtless use of words, are disproportionally blown out of context and interpreted as symptoms of oppression for large segments of the American people. Even though, they lack the larger context of the election circus that often lives off ostracizing average Americans.

American leaders, businesses and media describe many of the nation's most cherished historical figures and institutions as inherently bigoted, racist, and hateful of foreigners. They depict the socio-economic system of the U.S. as inherently unjust. This does not stay at home. Rather, it affects foreign affairs. Many foreigners then equivocate the United States to China, Russia, or Nazi Germany.

They hear of Nazis and racists dominating the streets and boardrooms, and politicians on both sides accused of being fascist. When being told about widespread animosity to foreigners, skin color and ethnicity dominating daily interactions, and actual or apparent extensive discrimination on the left and right, they lose track of all but one notion: something is rotten about America.

Foreigners notice how America's political discourse focuses on issues that seem strange and secondary to most people's lives. Often, they contradict many of their own experiences, cultural values, and principles. Ever more in the U.S. seems about words and abstract concepts, rather than problem-solving. This lets them question whether Americans really are as despicable as they claim about each other. Or, they wonder, is the U.S. simply a continuously dysfunctional soap opera dominated by lobbyists and without the ability to deliver effective government and solutions for the world?

And all the while, American economic growth shrinks, its debt levels explode, and regulations and bureaucracies expand. Civil liberties are being curtailed and only slowly restored by courts, if at all. The Patriot Act and anti-terrorism laws grant the government of the U.S. some capabilities and powers reminiscent of totalitarian societies. If you get delivered a National Security Letter or get caught up in an anti-terror dragnet, good luck to you, or at least to legal "due course."

In reality, and obvious to those visiting or living in the U.S., there are, of course, vast differences to genuinely autocratic societies. And many issues in most other democratic nations are similar and usually worse. But the distinctions often seem a matter of degree, no longer of principle.

The negative aspects of other democratic societies are not as aggressively advertised across the globe. Maybe they should be. Australia made encryption illegal unless the government gets a back door. New Zealand's government may force access to people's mobile phones.[106] The U.K., France, and Germany can curtail free speech if words contradict school curricula.[107] European politicians demand top-down commissions that determine which "truths" can be told in public and which ones not. Look at people opposing officially promoted priorities and positions, like specific climate-change or Covid-19 *policies*. Many get canceled, fired, or publicly censured.

Who instinctively thinks of Soviet-style Politkommissars, or Chinese communist ones, may only be off when comparing the extent. And most of the intrusive tools of American and other Western countries are today even more extreme and totalitarian than those that the Soviet, Stasi, or Nazi governments ever had. Or did these regimes know where almost every person and every car in their country was at any point in time and whom they interacted with? Did they have access to nearly all data generated at any time by all people, including almost all financial transactions?

Then arrived Covid-19. Restrictions of civil rights became extreme by *any* traditional standard. Most people justify this with special circumstances. Maybe they are right. But if so, where exactly are the boundaries for curtailing business activities, preventing people to gather, announcing curfews, and letting the old and sick die alone and without any of their families allowed to visit during their last and often painful moments?

Or what about limiting the freedom to travel internationally and even domestically, prescribing what to wear and how to interact with each other? What does it say when we curtail free speech and ask private gatekeepers like Facebook, Google, Apple, and Amazon to take part in enforcing all the above? So-called free societies made their population's daily activities dependent on the whim of an unelected technocratic elite empowered by small numbers of politicians. Those "in charge" seemed to manage things off-the-seat-of-their-pants and squashed alternative opinions. Then, they either ridiculed or attacked opposing thoughts and actions, and censored scientific and political debate.

Whether or not justified, the actions I just listed in the past handful of paragraphs eroded the image of a genuinely free society. It then is only a minor step to see differences to authoritarian regimes purely as a matter of degree and circumstance, and interpretation.

China does not automatically gain in popularity when the appeal of the U.S. drops. But with all that I just mentioned: "poof" goes our soft power.

7

Technology

Two Alphas or Two Underdogs?

Here I am participating in a foreign relations conference in Washington, D.C. Nodding heads of other attendees surround me, when a recently retired high-ranking general talks about the Chinese just stealing technology and copying weapons. I hear about America having the best military with the best weapons in the world, and the necessity to make sure that China cannot be Asia's or even the world's dominant power by 2049.

Yes, most seem concerned about 2049, not 2029, or today. It is the stuff that major strategic blunders and horrible wars are made of.

I hear diplomats and think tank analysts talk about America's global appeal. They urge the U.S. to continue to lead and remain the dominant power in Asia. And I wonder when these people step out of their virtual bubbles and into the real world. Based on what hard metrics are we the "dominant power" in Asia? What information is evidence for assertions the Chinese would (still) be mostly copycats?

The people making such claims are no dummies, in the opposite. They are extremely intelligent, well-traveled, and thoughtful professionals, diplomats, academics, and civic leaders. Still, they live in an echo chamber. I don't see them at technology conferences. How can they judge whether China just copies our technology? What makes them downplay the ingenuity of this Asian society, and indirectly so many other countries around the world? Do they know the numbers, and their trajectories, of patents in emerging technologies, from artificial intelligence to virtual reality and biotech?[108] Or who drives the digitization of our society using mobile fintech, or the distributed ledger and blockchain revolution?

The Old and New Multipolar World of Tech

A lot of technology leadership comes out of tiny Switzerland, Hong Kong, and cities like Berlin, London, Nairobi, and Singapore. Many pioneers work in Shanghai and other Chinese cities, universities, and corporate R&D labs. Typical science and engineering teams are spread all over the world. Chinese money funds startups everywhere and also offers public-private partnerships with access to near-unlimited resources, vast databases, and the most aggressively growing large market on Earth.

The U.S. still often plays a dominant role. I still see it overall ahead of other individual countries. But not across the board. Silicon Valley, Boston, and Austin no longer dominate the world's venture capital and technology sector as they used to. Neither do American companies. The picture quickly changes if you take out a mere handful of them. Without the Bay Area's Facebook, Apple, and Google, and Seattle-based Amazon and Microsoft, the world already looks much flatter. The big ones lead, alright, but also leverage global talent, international development teams, and offshore manufacturing capabilities.

To be fair, some in the U.S. government notice and try to change the direction. Just three weeks before the just mentioned conference in D.C., I attended an online seminar organized by the U.S. energy department and some of America's best universities and research institutions. The main presentation had the title "Can the US Compete in Basic Energy Sciences?"[109] Note that it did not use the words "dominate" or "lead," but "compete." The honest answer given was that yes, it is possible, but will require a major effort. It has been getting increasingly difficult because other large countries have been catching up in Europe and Asia. And China plays a huge role in it as well.

In the mid-19th century, it had been the United States that started on its path of ascendancy by copying and adapting other nation's technologies. It then enhanced and advanced them and developed an exports market. China has been on a similar path. Many of its businesses reverse-engineer and copy, even steal, and then they modify and enhance. Frequently they have been, and still are, assisted by governmental authorities and what seem to be coordinated paramilitary hacker groups.

But even the most aggressive ones find less to steal and copy because they have started to out-innovate their Western targets. And anyway, we in the West are not innocent victims of this. Much of Silicon Valley's success is rooted in cooperation and the free flow of ideas and knowledge. It focuses not on secrecy but on open source and sharing of knowledge. Likewise, Western businesses usually transferred their intellectual property voluntarily and in exchange for temporary access to Chinese markets and profits.

In the 2020s, this is changing. Profits are down. And most established market and technology leaders have gone as far as they can and the curve is flattening. Chinese competitors are taking over, and foreign corporations experience a much rougher ride in the Middle Kingdom.

Western entities are used to their home countries' slow-moving regulations and legal and lobbyist protections. These frequently use regulations to engineer politicized outcomes, like those related to environment, social, and governance (ESG). Meanwhile, their Chinese counterparts aggressively pursue their personal and their society's advancement through a highly effective Chinese-style version of crude capitalism that operates at high speeds and with relentless innovation cycles.

The copying of technology continues but is also becoming old news. Like so many things in the technology world, and like the U.S. itself, it seems to have followed a logistic curve. In the 1980s and 1990s, the level of education and experience of average workers in China was so low that the country had relatively little *ability* to absorb foreign intellectual property. For example, if you have not seen Western-style robotic manufacturing processes yourself, how can you contribute beyond being a mere robot-like contributor yourself?

By the beginning of the 2000s, though, China already graduated about three times as many engineers as the U.S. This was not yet a change of guard. There were some questions about these numbers and the quality of the education.[110] And even several years of high graduation rates cannot fully compensate for decades of advantages gained through high-quality education and subsequent professional experience.

In the mid-2010s, then, the World Economic Forum reported that in 2013, China's R&D expenditures had reached the total of the European Union. The awakening Asian giant graduated about eight times the number of STEM students of the U.S., and the total of doctoral degrees in Science & Engineering (S&E) approached the U.S.[111] According to other statistics shared by the American Institute of Physics (AIP), China had been ahead of the U.S. in the total number of S&E doctoral degrees earlier, since before the turn of the century.[112] The spread was even more pronounced when excluding temporary visa holders from the American numbers.

China now has mostly caught up with infrastructure projects for housing, roads, and power generation that took the U.S. over a century to build. Forbes reported a calculation made by Bill Gates showing that China consumed more concrete in the three years from 2011 to 2013 than the U.S. did throughout the whole 20th century.[113] In 2019, China produced and used more than half of all steel produced on Earth.[114] It still plans to bring online about one coal plant *per week* for about another decade.[115] This is for a country of similar geographical size to the United States. And there is more. China is blanketed with modern airports, high-speed rail systems, and manufacturing facilities for high-tech industrial products from e-vehicles, airplanes, and nuclear power plants. Foreign competitors are being kicked out or, through a combination of market and political pressure, slowly motivated to seek markets somewhere else.

China is confident that its engineers and research institutions can cope and take it from here. The country, now, is churning out not only many times the American numbers of engineering and STEM graduates, but also closes in on their quality. It would have been easy to be satisfied with remaining just an expanding contributor to global supply chains and keep accumulating wealth while consuming foreign consumer products. But China no longer sees this as necessary. It can manufacture more cheaply, often even near exclusively because competitors of Chinese companies also build their products in the country.[116]

With their cultural closeness, Chinese companies even can have advantages when designing for the local markets. They need no foul play or much of regulatory "help" to push out competitors. Much of it comes from consumers' preferences. If Apple generically designs and builds for the global market while cheaper Chinese smartphones specifically target the local market, Apple will lose local market share.

This shift happens in all industries, including high tech and its suppliers and derivatives. China makes hundreds of billions of dollars available for basic research in the fundamental, or "basic," research of deep technologies. "Deep" means they involve high scientific and engineering complexity and underpin other technologies. Low regulations then enable the best businesses to quickly commercialize high-tech products and innovations for over one billion people across the second largest market in the world.

Researchers, startups, and innovators keep flocking into the country to afford costly research and development. This is especially noticeable in the PRC's priority areas of quantum computing, artificial intelligence, aircraft manufacturing, space technologies, and also virtual reality, additive manufacturing, and biotech. And so, China's advanced technological capabilities and the volume of its exports keep expanding even while its dependency on foreign markets is shrinking, from 27% in 2010 to below 20% a decade later.[117]

Judging by a proclaimed lead of published patents, ever more original research originates in China.[118] Obviously, there are still structural issues. Academics and industries are not working as closely and effectively together as in the top U.S. educational institutions. Scientific advances benefit and often even require openness and collaboration, which is the antithesis to authoritarianism or even totalitarian control. But the country's leaders are aware of this and try to change it. They hint at, or outright offer, quick and unbureaucratic access to a market of over one billion middle-class consumers. Often, it is even more appealing just to get access to these consumers' data.

Today, when such data is the "new oil," this is a gigantic differentiator. It allows for entrepreneurial experimentation and for lots of AI-driven empirical research using near-real-time data from the most populous society on our planet.

As mentioned in the above chapter about economics, recent political centralization efforts cast some doubts on the PRC's ability to maintain this momentum. Also, other nations and groups strengthen their positions, whether Europe, Japan, India, and even outliers like the UAE. [119] Just looking at the two current superpowers, though, should make us alert. We can and should not bet on trend lines reversing themselves and China's efforts and recent advances failing.

Ignoring such trajectories would be foolish. We must get back into the game, regain our leadership, or at least re-establish our comparative advantage. To do that, we must think and act like a hungry challenger, not a comfortable hegemon trying to maintain a lead that no longer may exist. Or else, who do we want to be - infamous Kodak giving up its market domination and technological superiority because they could not get their priorities straight and change with the times?

AI, and Data as the "New Oil"

Fossil fuels enabled us to build the modern world as we know it. First it was coal, and then above all oil, that provided previously unheard-of dimensions of energy. By replacing animal and human muscles with machine power, it freed up time and enabled the monumental shifts of our workplaces. First, this moved people from the fields to the factories, and then from the factories to the offices. This eventually pulled almost all of humanity out of poverty. Around the world, people could live substantially longer, healthier and more comfortable lives, better educated than ever before.

Oil and gas additionally delivered our fertilizers, and its molecules are part of the food most of us eat. It is in the paved roads we drive on and the tires that do so, and in the plastics and paints and many more products we use every day. For all these reasons, fossil fuels enabled our modern civilization and nearly all the accomplishments of the past 150 years, from the industrial revolution to the computing revolution and the subsequent emergence of the Internet.

This is changing, although not mainly because of new sources of energy. Rather, future progress is driven by data more so than energy. Fossil fuels still provide roughly 70% of the world's energy and will remain similarly dominant for at least another decade or two.[120] It is the reason nations are concerned about their supply of fossil fuels and navies keep being important to protecting it. But although digitization needs energy as well, this is not the crucial ingredient limiting or enabling its success. What the digitized world of our near future needs even more urgently is data.

Data is the oil of the 21st century. Whoever owns and uses it can predict, optimize, or manipulate. The tool that does so is called machine learning, a subset of artificial intelligence. It finds patterns where humans cannot and does so at lightning speed. This then enables us to move beyond mere automation and create autonomous machines that react based on environmental stimuli. It increases our abilities to manufacture, to make or keep us healthy, and to improve nearly every aspect of our society and lives.

This has not only profound implications for the future of humanity. It also affects international relations and threatens a profound upheaval of the global system that we have become so comfortable with. Henry Kissinger, Eric Schmidt, and Daniel Huttenlocher point this out in their before-mentioned book "The Age of AI." They correctly point out AI's vast powers.

However, only focusing on AI, as they did, would be too narrow. Many technologies keep developing at accelerating speed and create opportunities and dangers for international relations. These technologies can decide wars and win peace, make us dysfunctional or prosper. They can enslave us or set us free.

Exponential Transformations

This may sound like hyperbole or at least a step into the far-off future, but it is not. We must keep in mind that exponential rates of tech-driven change result in mutually re-enforcing and continuously accelerating trajectories. It is not only Moore's Law, named so because Intel's cofounder Gordon Moore observed an approximate doubling of computer processing power, roughly every couple of years. This goes beyond computing power. Several other powerful technologies continue to grow exponentially.[121]

They include, for example, communication speed, data storage, and sensors. Together, these drive advances that often are accelerating even *more rapidly than Moore's law*. We can see some of that in artificial intelligence, genetics, biotech, age-reversal, nanofabrication, materials development, and robotics. A couple of decades ago, inventor and futurist Ray Kurzweil described the effect in a seminal article and called it "law of accelerating returns." To date, his projections have mostly been proven right.

It is not natural for humans to understand what the tech world labels as *exponential* growth. We are comfortable thinking in *linear* terms of small percentages, not in doublings. And this makes us grossly underestimate the transformative power of change.

For example, some of the just mentioned technologies double their capabilities every two years. After 10 years, this will not mean that they are five times as strong, which would already reflect extremely rapid growth, although a "merely" linear one. Instead, it would be 32 times.

We may even better appreciate the profound impact of these numbers when we translate them into the world of economics. After 20 years, linear growth of 10% delivers an economy almost eight times as large, while after the same two decades of bi-annual doubling it would be over 1000 times!

This kind of growth is not theoretical. It has happened for many decades already in the world of computing, with similar growth rates of processing power being projected for at least another two decades. It is the main reason the digitization of all aspects of society advances at exponential rates.

While we should be careful to project the past into the future, in these matters we have outstanding empirical, mathematical, and data-related evidence that the growth will remain exponential for quite a while.[122] And again, we are not just talking about one technology but many of them, nearly all of which are critical to human advances.

But some of these advances, specifically those involving AI, are on a trajectory multiple times and even several *orders of magnitude faster*. If AIs would double their abilities at the above-mentioned rate, every two years, a decade from today they would be 32 times as powerful. If each doubling would only take one year, AIs in 2032 would be one thousand times as capable as today's. In reality, though, some aspects of artificial intelligence currently *advance at rates of 10 in about one year*. The continuation of this progress seems likely for at least another few years. Over a decade, this would increase the power of AI *ten billion times*. In 12 years from now, so at about 2024 or 2025, AI's would then be **one trillion times as powerful as today.**

Do we want China to have these capabilities before we do? And if we are first, how do we want to use these powers? Do we trust our governmental or business elites enough to wield them responsibly?

China's 1.4 billion people generate more data than any other society. They do not even pretend to have privacy. The PRC's political system makes it easy to centrally collect and use all data generated on the Chinese mainland. This means that Chinese tech giants Baidu, Alibaba, and Tencent have access to more oil of the future than anybody else.

And so does everybody who works with them.

Web3 and a Future Rules-Based Liberal World

The just described changes are not optional. We cannot wish them away. They *will* come, as plenty of empirical evidence shows.[123] Check out https://techtonicedge.com for updates on the breathtaking pace of technological change in a variety of highly relevant fields.

But I still frequently encounter critical arguments that the above description of technology-induced transformations "does not make sense" and would therefore not become reality. For many, the changes seem too far-fetched and remote from the realities of our biological needs and comfortable daily routines. Others acknowledge today's capabilities of new tech but understandably struggle with the just described natural limitations of humans to grasp exponential change. They fail to appreciate what the multiplication of data collection and processing capabilities by factors of 10, 1000, 1 million, or 1 trillion can or will mean for our near-term future. Again others are concerned on a deeper philosophical level about what this means for our humanity.

And so, although they sense that technological changes will upend all aspects of society as we know it, far too many people hear the words but decide to ignore them. They choose to assume, or pretend, that there only will be linear and incremental changes. Then, they hope the cozy old will prevail and the world will "go back to normal again."

But when I point at these underlying and truly transformative changes, this is not me describing what I wish *should* happen. It is me relating the reality of what *is* happening and *continues* to happen, measurably. People and nations *will* take advantage of such tech-induced new opportunities. Those that do will in the end be better off and more prosperous and powerful for it. Most likely not just "better" and "more" so than ever before, but also when compared to those that ignore the inevitable.

For better or worse, almost all people around the world spend ever more time in the digital world. Social media define communities, online courses educate, scientists, engineers, and businesspeople cooperate virtually, and companies sell online. Most global commerce is conducted digitally, and by 2022 at least 60% of the global GDP will be digitized.[124]

But complete digitization means more than using computers and software, creating websites, buying online, or using Twitter and Zoom. It is more profound and reflects real life in what we nowadays call "metaverses." Objects, machines, and even humans get "digital twins" that are in real-time connected, using sensors, to their physical embodiments. This is already happening at a large scale today and permeates many industries. The so-called Internet of Things and Industrial Internet of Things (IIoT) consist of tens of billions of sensors, soon to be a trillion.

They connect and combine the data they collect with those from our activities online and on our personal computers and phones. This then becomes the data trove that advances in AI are based on. Today, almost all these data are owned by big corporations, in the West by no more than a handful of them. Our challenge is to securely make them available as a public good and introduce private ownership and competition. This empowers people to use the data for innovative purposes that benefit them and society, not just a handful of well-connected massive enterprises with limited aspirations. Today, it is as if the oceans would be owned by companies. Facebook owns the Indian Ocean, Google the Pacific, Microsoft the Atlantic, and Amazon all the ports. We can use them, but they can centrally direct and control our activities.

All the capabilities I mention here are already in use today and being massively scaled out. Ownership of data is created, secured, and transferred on top of such technical platforms and within the metaworlds created on them. This new digital infrastructure is called "Web3" (or sometimes also: Web 3, or Web 3.0). The first generation, Web1, was all about creating web sites to inform, advertise, or sell. The second generation, Web2, enabled the interactive participation of users, via social media, chats, forums, and similar means.

Web3 is about ownership and commerce. The just-described digital twins replicate products, cars, devices, and even humans. Often, products will even be custom-designed and already created via borderless virtual collaborations. Their sensors will enable remote teams, or even AI, to control or maintain the physical objects, whether these are inanimate or human bodies. And then, these physical objects, or their services, will be traded on these platforms. These will be designs for cars, software, even physical real estate.

The overall effect is like when enterprise resource planning (ERP) software entered the business world around 1990 and broke down the barriers *inside* organizations. Managers got near-total visibility. Suddenly, they could run companies much more efficiently. Now, Web3 eliminates the barriers *between* organizations and people, without gatekeepers and intermediaries. It is about to link the trillion sensors of the Internet of Things, enabling and recording everything about anybody and anything of value. This capability makes Web3 absolutely critical for the future of nearly all commercial and human interactions.

The big question then becomes whether this global commercial infrastructure should be centrally controlled, as obviously envisioned by China and others that trust top-heavy approaches more than the people. Or should it trust the rules-based self-organizing powers of the masses? The latter would be bottom up and therefore protect and empower individuals and their rights and freedoms.

Because data is becoming the new oil (or "lifeblood") of society, no genuinely free and liberal society can exist in the future unless the rules of Web3 (1) secure all data and communication, (2) extend legal ownership rights to people and entities operating inside Web3, (3) safeguard their identities, and (4) protect the commerce performed by them.

This makes it essential how we organize, secure, and manage the virtual platforms and worlds of the emerging Web3. If our goal is a global rules-based liberal system, then the digital world and Web3 must be owned by the people and remain free from corporate, criminal, and government control.

Part II

GRAND DELUSIONS

Fighting Great Power War in the 2020s
Abroad and At Home

8

The Trap

Exponential Change and the Future

"Right, as the world goes, is only in question between equals in power, while the strong do what they can and the weak suffer what they must."

Thucydides (Melian Dialogue - ca. 300 BC)[125]

"So what?"

"We are still the world's number one."

I frequently hear this even when talking or listening to experts on international relations and members of the U.S. foreign policy community. Politicians and the media also keep making similar statements. They repeat these words, often reflexively. It is therefore not surprising that the public, and even otherwise well-informed people, don't seem to acknowledge that something serious is changing. They hear that the United States still has the world's largest economy (by nominal GDP), best technology (Apple,

Google, Amazon, Microsoft, Tesla, and SpaceX), and above all the most powerful military (with 800 bases worldwide).[126] It matters little, they say, that China is going to eventually outgrow us.

And look at all those aircraft carriers, fighter planes, drones, satellites, and what have you! And our alliances! We can prevent the Chinese from dominating East Asia and the world. We keep outspending all our military opponents taken together. It therefore just takes resolve and marching orders, and our military will deal with any threat. This may become ever more costly, but perhaps the proper conclusion is to fight sooner rather than later. We don't want to end up being too late, like Chamberlain in Munich, when confronting Hitler in 1938.

As the indispensable leaders and defenders of the free world, many even insist, we have an *obligation* to confront China's authoritarianism. Only variations of deterrence or containment can reliably prevent the China menace from growing to a level that eventually threatens the American mainland. We therefore must be "resolute" and "deny" China any change to the status quo.

If it just were so simple.

Sprinkled with formally correct factoids, the above points all *seem* correct. They form the basis for most of the smart suggestions that many well-informed and intelligent people keep making on how to confront China. And still, the overall picture they paint ignores some highly relevant developments. Above all, we are way past the point of preventing a shift in global power relations and denying China a dominant position in East and Southeast Asia. As I described in Part I of this book, technology, demographics, and economics already have taken care of this.

But while most experts superficially pay tribute to many of these changing dynamics, their dominating instinct is to continue discarding, ignoring, and spinning uncomfortable evidence. This includes East Asia's profound regional military imbalance in China's favor, local cultural and historical factors, and the actual economic and soft power structure in the region. Above all, and beyond regional points, "we are still the number one" is mostly about summary numbers. It does not sufficiently consider the exponential effect of economic trends. And it neither takes into consideration the profound repercussions of new digital and space technologies and how they threaten the critical infrastructure of America's heartland. Today, the United States and its citizens are indeed more vulnerable than at any other time in the nation's history.

Most see our digital capabilities as an advantage, with good arguments. But this only is the case in "normal" times and much less so during war. Digitization in a networked world introduces new dependencies. Interdependent nodes risk cascading failures. This threatens to devastate key national infrastructure simply by taking down a few core elements. Most parts of our critical infrastructure are hierarchically organized or dependent on a relatively small number of critical nodes. And sometimes it is not even necessary to take out specific ones. Take, for example, our electricity grid. It operates so close to the margins to maximize efficiency that taking out enough components can cause large parts of the system to collapse.

Ignoring the implications of this is dangerous and lets us set the wrong priorities. We correctly identify and analyze part of the situation (the huge capabilities of our networked world), but not how its underlying structures affect our security (how easy it is to bring it down). It is a bit like focusing through a sniper rifle's sights on a lone infantry soldier sneaking up on us, while not noticing the tank across the street. Academia, think tanks, journalism, and even politics often promote theories. These frequently ignore the larger and more complex and consequential parts of the picture. Everything goes when you don't have to prove your point in battle. Until it is too late, of course, and the fight turns real.

And so, the Beltway and its outposts count and count, from military budgets to the number of platforms, global bases, and the aggregated economic output of our allies.[127] When the totals or actual capabilities don't add up anymore, particularly today in the Western Pacific regional theater, thinkers, writers, and politicians then bring on some magic. They then use terms like "quality," "high-tech," and "experience." If even this becomes too hard to defend, they add belief and moral imperatives.

At that point, we hear about our solemn duty to defend anybody who lives in societies whose institutions resemble ours. This comes with faith in our success because of a supposed indisputable superiority of America's economic and political system. It does not even seem to matter to what degree we and our institutions have already compromised the supposedly universal principles and systems we allegedly hold so dear. Or how many times in the past we failed to live up to comparable commitments. This also brushes aside how many of America's self-absorbed and self-incriminating domestic political and cultural battles and priorities weaken our hard power and reduce our soft-power appeal to others.

What This Means for the Western Pacific

Applying this to the Western Pacific theater, where we would have to confront China if we were to militarily respond to an attack on Taiwan, we quickly encounter discrepancies and contradictions. The numbers of many Chinese military platforms, increasingly even advanced fighter jets and ships, approach or outnumber those of the U.S. and its allies. In some fields, the PLA has more missiles than the U.S. has realistic defensive capabilities. This makes the U.S. Navy susceptible to saturation attacks. Other missiles are so fast and maneuverable, like the DF-17, DF-21, or DF-26, that the kinetic energy they release may in theory even be able to sink a large ship without requiring a warhead.[128]

Many of these missiles have no equivalent on the U.S. side. Others far exceed the range of the American systems they target, especially its fighter planes. Combined, this means that the Chinese can strike at the U.S. Navy and American bases with near impunity while the U.S. military cannot even effectively retaliate with its most potent conventional weapons. And the Chinese People's Liberation Army, like the Russian military, has also developed novel weapon systems. Among them are ballistic and hypersonic missiles that the United States cannot defend itself against - even in North America.

The latter point is particularly notable. Chairman of the U.S. Joint Chiefs of Staff General Mark Milley's characterization calling a Chinese hypersonic missile test a "Sputnik moment" may be a tad sensationalist. We also have somewhat comparable capabilities in similar states of readiness.[129] Such a demonstration alone does not require us to completely change our defense and security priorities as the Soviet Union's launch of its Sputnik satellite did in the 1950s. *But* it still is an important additional data point contradicting the fairy tale of the Chinese relying solely on espionage and theft instead of innovating themselves.

Hypersonic gliders require advanced missile or even space technologies, sensors, and targeting mechanisms. To produce them in high enough numbers and establish an overwhelming arsenal, they also need complex manufacturing competencies. And such scalable high-tech capabilities using advanced manufacturing prowess sounds close to a definition of China in the 2020s.

As described in chapters 5 and 7, many manufacturing and technology companies in China operate in a highly competitive crude capitalist environment. The winners take home giant prizes by getting access to a vast billion-people market - and to these people's data. And economies of scale almost immediately kick in. For modern tech products, particularly those supported by AI and requiring massive amounts of data, this is huge. And then, state-owned companies and the government are ready to support and leverage this right away for military or paramilitary purposes.

While proving all of this in the civilian world, it is also how the PRC, at least since President Xi took office, tries to develop and deploy new weapon systems.[130] It is not so much, though, that China is, or has to be, more successful in any specific part of this process. But its focus is much narrower and centers on *regional* superiority that specifically targets the U.S.' capabilities. Like in business, innovation often can deliver a bigger punch when catering to local markets versus the competition's global and therefore much more "shotgun-like" approach. With this context, it matters that fighting a genuine peer power in its home territory is something the U.S. never has done to a comparable degree. Japan was a military peer power in the early 1940s, but it was a comparable economic lightweight.

The Tech Superiority Fallacy

Recent developments of weapon systems suggest that China's PLA is closer to Silicon Valley's hyper-iterative rapid innovation cycles than the Pentagon.[131] It leverages concepts of rapid iterations and continuous improvement, otherwise perfected by American capitalism. This can lead to seemingly inefficient development efforts early on. But it also establishes a high velocity with positive feedback loops that effectively drive innovation. The U.S. DoD, in contrast, mostly sticks to old school centralized planning of large weapon systems lasting generations. Mind that this does not refer to a *technology* generation, which nowadays typically is quantified in months, but a *human* generation measured in decades.

A bit of personal experience puts this into perspective. In the mid-2000s, I participated in a symposium in Singapore about complex project management. This is a key tool used, among others, by the U.S. military to develop highly sophisticated weapons platforms. At its core is a "system of systems" approach to manage and coordinate a highly complicated design, procurement, manufacturing, and validation process. One participant had just recently been tasked to manage multiple billions of US dollars to advance development of the Joint Strike Fighter (JSF), nowadays better known as the F-35. This was about 15 years ago, and only recently were finished planes put in service. When I was at this multi-day meeting in Singapore, my youngest son had just entered elementary school. By the time the first functional fighters were fully ready to deploy, he had graduated from college.

Is this supposed to be normal? In the meantime, we saw the emergence of Facebook, Tesla, Bitcoin, SpaceX, Instagram, Stripe, Twitter, and more than a dozen versions of an iPhone that had not been invented yet when I was in these meetings. And we also witnessed the rise of Chinese tech success stories like Xiaomi, ByteDance, Didi Chuxing, and Ant Financial.

Mega-projects like the F-35 fighter plane are futuristic and a technology generation ahead when being designed - but in many ways outdated and behind the commercial world when put in service 10, 15, or 20 years later. They outline genuinely forward-looking tech solutions like "networked situational awareness" for F-35 fifth-generation fighter planes. But when they finally get to produce the first planes more than a decade on, some of its "high-tech" components have fallen behind what commercial industries produce for mobile phones, industrial robots, and self-driving cars. And increasingly, such commercial innovations originate outside the U.S. as well. A combination of slashed research budgets, ever more restrictive regulatory environments, and risk-aversion make sure of that.

In contrast to military procurement cycles, the tech world releases every day new virtual reality capabilities, better autonomous or semi-autonomous systems for driving and flying, new materials, and more advanced medical capabilities. Companies like Tesla modernize their cars *continuously*, with new updates and improvements available for download every few weeks. These deliver quality boosts driven by the crowdsourced sensory input from millions of cars or devices, optimized by AI.

In a riveting book called "Kill Chain," Christian Brose compares these contrasting approaches in chilling detail.[132] As the late U.S. Senator John McCain's former principal advisor on defense and security, and later his Chief of Staff, Brose is a Washington insider and defense expert. He has what he calls "access to the Pentagon's most highly classified secrets and programs" and regularly "met with our nation's top defense officials and highest-ranking military officers."

His detailed references to the state and trajectory of emerging technologies reflect a keen understanding of what is going on outside the DoD ecosystem and beyond the Beltway. But they also make clear how far we are behind in our all-too-comfortable approaches and how our institutional limitations and political processes waste money and set the wrong priorities. The latter points are particularly eye-opening, explained by an obvious expert with personal experience on a high level of politics and government.

Since the 1960s, our technological and economic superiority seemed so obviously true that we still treat it like a first principle instead of the ever more likely truism it has become. How do these apparent experts know? I don't see many U.S. military or foreign policy experts from think tanks at emerging technology conferences.

How can one be a genuine expert, though, without diving deep into the most transformative developments affecting all other aspects of our lives? Where is the profound knowledge in the institutions and among our leaders, about the digitization of near-everything, artificial intelligence, quantum computing, distributed ledger technologies, 5G and 6G, additive manufacturing, virtual and augmented reality, brain-computer-interfaces, genomics, and low-earth orbit satellite constellations? And where is then the actual experimentation and deployment of rapidly iterated weapons systems and solutions?

This is not a blanket criticism. Indeed, many political advisors and the DoD work with academics who know what goes on in the commercial world. I also just recently took part in a workshop about the future of exponential technologies that included members from the military. Many "get" it. The U.S. Department of Defense funds basic research, even lots of it.[133] It attempts to drive defense innovation through the Defense Advanced Research Projects Agency (DARPA) or the Air Force Research Laboratory (AFRL). To "strengthen our national security by accelerating the adoption of commercial technology throughout the military and growing the national security innovation base," the Defense Innovation Unit (DIU) focuses on the right kinds of technologies.[134] The government even has venture capital arms like In-Q-Tel.[135]

This comes with a big "but," though. And the resistance to the U.S. Marines' *Force Design 2030* modernization plan is exemplary for it. Devised by the men and women who actually do the fighting and informed by an acute understanding of technology and its trajectory, it advocates for a profound redesign of this branch of the

American military. As a result, as the former Marines in U.S. Congress that support this plan state, "it shifts resources from armor, artillery and manned aircraft to invest in longer-range missiles and unmanned aerial systems, better sensors and surveillance, and the development of new cyber forces."[136] The opposition to the plan is huge, although in principle it almost perfectly captures the essence of upcoming shifts and is therefore exactly the right thing to do.

This is consistent with our experiences during the past decades. Many people making the most critical decisions and allocating budgets seem to be constantly out-of-tune. The defense budget continues the acquisition of large platforms designed and built with programs sometimes lasting for decades, delivered by only a handful of vendors. The slow turnover and lacking competitiveness of these processes are far behind the accelerating innovation cycles of commercial industries.

Another look at fighter planes makes my point. Computer chips, sensors, and AI-assisted data communication protocols and algorithms are at the core of an F-35's system. They create situational awareness and facilitate fast decision making for the most effective deployment of weapons systems. However, do generals, admirals, and politicians really believe that, as good as they may be, Harris-built integrated core processors awarded in 2018 and put in service in 2023 into *hundreds* of flying machines are superior to those spit out in a highly competitive environment via annual iterations for *hundreds of millions* of customers, like Apple's M2 or A16 Bionic Chips?[137] Or that the F-35's computing systems could be superior to NVIDIA's Grace™ or A100 products? Or that their artificial intelligence capabilities could beat Baidu's, Tesla's, or Alphabet's AI algorithms?[138]

The latter power highly competitive applications for tens of billions and soon trillions of devices in highly complex and fluid environments. These reach from speech recognition to autonomous driving or flying, and far beyond. Even without knowing more than these numbers, Wright's Law would let me bet on the superiority of civilian products. It states that high-volume manufacturing builds experience, drives quality improvements, and therefore accelerates technological progress.[139]

"Military grade" used to be a distinguishing term for quality. Nowadays, though, it often seems to mean "does not as easily break when falling down or being exposed to dust." Ask an operator about some of the gear he gets, and how far they are behind what some in private industry use. Or the guys and gals sitting in containers, manually trying to make sense of massive amounts of non-integrated data feeds from disparate systems. How do the tools in their day job stack up to the crowdsourced AI-driven autonomous Waze app they use to find the fastest way home from base?

Of course, the situational awareness of a Tesla on autopilot is far from perfect. But it also deals with a fluid and complex environment and massive inflow of data from sensors. How, then, does this $50,000 machine stack up to a $50 million plane that is part of a $50 billion naval strike force?

It seems more probable to me that our fighting against technologically far inferior opponents for decade after decade has made us complacent. For too many of us, it is easier to stay in our bubbles, often without realizing it. As we all know, search engine and social media algorithms distort our perception and confirm our biases to optimize our comfort level. Sensationalizing media and news cycles in a hyper-politicized environment cause harm as well. They discredit many of those thinking independently and challenging poll-driven supposed "truths" of never-ending political campaigns.

Therefore, like the politicians and the DoD with their antiquated procurement cycles and systems, many of their advisor and thinkers develop tunnel vision. For social and political self-preservation, they stick to outdated ideas. Many may even believe that what they say is reality. And then, they feel overly confident about still recommending the same kinds of weapon systems and strategies in the 2020s that successfully beat Hirohito's and Hitler's forces in the 1940s: **aircraft carriers and manned fighter planes!**

Nobody Remembers, Nobody Knows

Barely more than a handful of people alive have actual battle experience in great power war. So, what do any of us really know about how capable or not our military is when it counts on the really big stage? Skills gained in Fallujah and Helmand Province matter a lot when fighting in similar circumstances, like the clearing of buildings and bombing of caves and small vehicles. All this counts much less, though, when confronting large numbers of troops who enjoy air superiority and are equipped with sophisticated high-tech missiles, drones, and other weapons that in several ways are superior to our own.

We may accept the unproven premise that we have the best-equipped and most powerful military machine of all times that eventually still can beat any opposing force. But this is highly unlikely going to be the situation our forces would have to deal with. In East Asia, our military would fight a war in a region where the PLA has far superior numbers. It would be in a way that Chinese weapons were specifically designed for, unlike ours. China can target American regional bases and track carrier fleets, saturate their defenses, deny their access to the battlefield, intercept their (very long) supply lines, and attack U.S. forces from the Asian mainland with near impunity.

We then can retrench far from China's shores and contemplate a large-scale counterattack. But attack on what, the Chinese mainland? Or on by then fortified positions close to it, maybe even around or on Taiwan itself where the PLA may have established beachheads? And attack from where, being removed from our own homeland by thousands of miles of exposed supply lines? How would we do that? Where would we stage our forces? What alternative weapons systems and strategies would we use? To achieve exactly what? And in what time frame?

Alternatively, we can try to break the above limitations imposed by Chinese defensive lines and the PLA's superior numbers and capabilities of weaponry. We can decide to escalate and retaliate by attacking Chinese bases on the mainland and the launch pads and command-and-control nerve centers of their forces. Our forces can actively intercept their supply lines through precision bombing campaigns near or in the middle of wide swaths of the Chinese industrial heartland and population centers. Or we can establish economic pressure through a naval blockade.

However, China can match any of our moves. Step for step, tit for tat. They even can directly attack and harm the American heartland and most of the American population.

And this is exactly what likely would happen and where we would end up. Unless we leave it at one-sided battles on the open seas, there is hardly any chance either side could win such a conflict while keeping it contained. If the term "winning" even has any more meaning here than when used to describe the World War I carnage on a hill in the south of the Alsace.

Nearly without exception, all recent simulations and war games point exactly at that outcome. But I posit that the picture usually painted by such dire results is still too rosy. If we were to fight now, we would not just lose the war in Asia itself. We would *also* risk incurring mass casualties and a stupendous scale of destruction on the American continent. Escalation is highly likely. *We* would escalate if attacked, and did so twice, in 1941 and 2001. And likewise, we would have to assume that hitting Chinese bases in their heartland would trigger Chinese retaliation as well.

Since World War II, the U.S. has only once directly fought a major power. It was in Korea in the 1950s. This power was China, and it did not end well. Hundreds of thousands of comparatively ill-equipped soldiers of the PLA caused tens of thousands of casualties for the U.N. forces under American leadership.[140] It resulted in one of the U.S. military's worst defeats and forced the longest retreat of American history, all the way from the Yalu River back to the 38th parallel.[141] Since then, while the U.S. has won almost all significant battles in every one of its subsequent wars, each one of its enemies had been vastly inferior technologically, financially, and economically. And even then, America rarely had the stomach to successfully pull through in the long run. It won the battles, but not the wars. The recent withdrawal from Afghanistan was just one example, and an extremely visible one at that.

Our post-World War II enemies did not fight like the Germans or Japanese in the 1940s. Often, they simply did not have the means. With far inferior capabilities, they used asymmetric warfare. On traditional battlefields, American enemies in Vietnam, Granada, Panama, Afghanistan, North Africa, and Iraq were no match. One can even question whether the China fighting us in the 1950s really was a great power and peer, considering its economic and technological backwardness. And still it achieved most of its objectives against the United States.

Comparisons since then have been theoretical. Calling the U.S. military the mightiest fighting force in the world and expecting it to beat China near its homeland is speculative, and more likely intellectually sloppy or insincere as well. A closer look at Russia's logistical challenges in the Ukraine should be a warning sign of what American forces may have to deal with. It would not even be next door but 6,000 miles from home. And not against a much smaller and weaker nation but one that is larger, more populous, and as close to an economic and technological equal as they come.

I consider the U.S. military one of the most trustworthy and professional institutions in America. But facts are facts. When we ignore them, it is at our best's peril and death. We then risk civilian lives and large-scale destruction of our society's infrastructure as well.

Beware What You Wish For: Simple Geopolitics

Perspectives also matter.

In the West, arguments for deterrence and containment of China sometimes go as far as asking for a solid commitment to militarily defend Taiwan. The most extreme positions even ask to let Taiwan acquire nuclear weapons to deter Chinese aggression.[142] It is classic peace-through-strength reasoning.

Even on a continent far from home, such a position still is tremendously appealing to the only power on Earth ever having experienced a "unipolar moment." But I have started this chapter with a quote from Thucydides' Melian Dialogues. Its logic is simple: on the battlefield, raw power matters more than anything else. Indeed, if you are militarily strong, then you are more secure than otherwise. You can defend yourself or overcome and even dominate potential challengers and enemies in your neighborhood and far away.

However, such statements do not say what the words "power" and "strength" mean precisely. And even more critically, they do not clarify who is more powerful in any specific situation. Throughout history, most societies that willingly engaged in war probably assumed that they had the advantage. Sometimes they were right, and the answer was clear. Frequently it was not. Then societies fought to find out, even over secondary and small stakes.

When one major power is ascendant and seems to be on the pathway to supplant the reigning hegemon, globally or regionally, it almost invariably leaves the question more frequently unanswered. This is when we hear about the "Thucydides Trap" that Graham Allison's project at Harvard analyzed.[143] It tried to learn lessons from the war between ascendant militaristic Sparta and the regional hegemon of Athens. By the way, the more democratic and for a long time commercially more successful Athenians found out the hard way who were stronger.

They lost.

The proponents of the peace-through-strength argument do not oppose a country's desire to dominate its neighborhood militarily. Far from it. They just are against *other* countries doing so.

Sometimes they don't care, for example, when these other countries are allied or far away.

And so, it does not bother the United States that Australia deliberately tries, and usually succeeds, dominating its Pacific Island neighbors.[144] The U.S. does not mind South Africa throwing its weight around on the African Southern Cone, or Germany calling the shots in much of its neighborhood. None of these actions are a serious concern, as long as interests are aligned and they do not challenge American core interests, militarily or otherwise.[145]

Arguments of process also matter, of course. On the Crimea, for example, Russia had and still has some substantive arguments going in its favor. Khrushchev somewhat arbitrarily moved control over the peninsula from Russia to Ukraine. Both countries also have a lot of common history, which blurs the picture and opens the door for many logically sounding arguments. Most of the Crimean population seems to support its belonging to Russia.

Even so, invading another country just does not feel right, and isn't! It also sets a dangerous precedent. The correct solution would be to hope the Ukrainians would let the democratic process work itself out locally, have a vote, and then implement the will of the people. All this sounds logical and good.

This is something, though, that almost never worked that way. Rarely anywhere in the world did a region ever manage this. Scotland did not, in the United Kingdom, Southern Tyrol in Italy, Catalonia in Spain, or the Confederates in the United States. The German Saarland succeeded, after World War II, aided by a French-German alignment that enabled the peaceful transition of its territory back to Germany after a popular vote. It is a helpful clue that alignment was the key to this successful transition.

The reality in nearly all parts of the world is that the shape of most countries usually is determined by force and then locked in until the next war or civil war. Looking at it this way makes it possible to consider the perspective of the other side, the supposed (or actual) aggressor. Countries surrender territory when they lose wars, but rarely get it handed back even after prolonged periods of peace. Like Mexico in the 19th century, Turkey after World War I, Poland after World War II, Germany after both World Wars, Russia after the Cold War, and China during the past decades.

From the Chinese mainland, therefore, things look different than from many academic circles and political power centers of the West. Foreign powers had been encroaching on Chinese territory and autonomy since the First Opium War started in 1839. The

Chinese empire lost small patches of territory to Western powers. Initially, the overwhelming majority of the Chinese population was unaffected until well into the 20th century. Simply stated, they were powerless and poor, anyway. But then, the entire empire crumbled, and imperial Japan started invading and brutally conquering extensive areas. Formally until 1916,[146] China was an empire and not really a Western-style nation state. The concept of a "nation state" is a relatively recent and somewhat nebulous European invention, less than 400 years old.[147]

Because of its immense size and population numbers, China usually dominated its neighborhood as the "Middle Kingdom," or "Central Kingdom." For centuries, it was a self-absorbed society. This changed with the integration of the Chinese economy into the global one since the 1990s. From that point in time, worldwide trade and shipping routes mattered to the PRC like never before.

Like when the U.S. constituted itself as a continent-wide hegemon and global power, China indirectly and then ever more directly began to dominate its neighborhood again, to various degrees. This time, formerly peripheral islands and territories mattered much more because self-absorption no longer was the primary source of strength. Rather, foreign trade had become important, as part of global manufacturing supply chains. Until the 2020s, this required above all access to and use of naval shipping lanes. They provided the crucial means to generate wealth and pathways to project power regionally and globally.

During its recent and still continuing rise, China looks at geopolitics similar to other great powers while they ascended. With a sense of continuing historic injustices, it used its newfound economic wealth and technological prowess to become a formidable military power.

Regionally, this turned the PLA into the dominant force near the East and Southeast Asian mainland's shores. But even globally, it enables China already today to confront America in substantive ways. Like other powerful nations, the PRC has not just the ability to defend itself near its borders but also to strike directly at the American heartland and its civilian population. It does not even need nuclear weapons for either of these actions.

This makes it important to look at the "why" and then at the "how" of such a rivalry spinning out of control, possibly ending up in a large-scale war.

On Cruise Control... Like It's 1914

Many of today's debates about Chinese geopolitical intentions and plans on Taiwan ignore or discount the implications of what I have so far laid out. I will summarize it in the following three critical points.

First, most foreign policy experts agree that *China intents to take Taiwan at almost any cost*, even if by military force. There are several reasons, but the core argument is geopolitical. Slightly simplified, it says that control over Taiwan and the adjoining islands is the only realistic way for China to break the geopolitical constraints that the First Island Chain imposes on its navy. These underlying Chinese claims pass enough of a low threshold to formally, historically, and culturally legitimize an attack on Taiwan in the eyes of almost everybody inside China (as far as we can tell in a society with totalitarian control over news and communications), and also many outside China. All factions of the Chinese political system have explicitly expressed this reasoning. It apparently enjoys enormous support in its highly nationalistic public.

Apart from the "how," the most significant remaining question seems about timing. Some pick symbolic numbers like 2026 as the 100-year anniversary of the founding of the Chinese Communist Party. [148] They point at a particularly appealing tactical constellation during the next couple of years or at President Xi's age.[149] Others see China waiting a few more years to take advantage of its still continuing trajectory toward global economic and technological superiority while it also establishes genuine nuclear second-strike capabilities, puts hypersonic missiles in service, and redirects supply chains for energy and raw materials.

Second, most military experts agree that *China has decisive military superiority* in the likely theater, and that there is no realistic way for the United States and its allies to win a straight fight so close to the Chinese mainland.[150] As just one of many examples, there is little to add to the following statement in the U.S. Army's magazine "Military Review" by Major Christopher Mihal in July 2021 about the PRC's Rocket Force:

> *"The conventional arm of the PLARF is the largest ground-based missile force in the world, with over 2,200 conventionally armed ballistic and cruise missiles and with enough antiship missiles to attack every U.S. surface combatant vessel in the South China*

Sea with enough firepower to overcome each ship's missile defense."[151]

It is even questionable whether the people of Taiwan would put up a sufficient fight, judging by its defense budget just barely hovering above 2% of the GDP. Taiwanese reserve forces attend military "training classes" measured in mere weeks.[152] Maybe the fight of the Ukrainian people against Russia's aggression will motivate them, or maybe its high cost in life and property achieves exactly the opposite. Although Taiwan has highly favorable mountainous territory with defendable coastlines, it neither has the same manpower as the Ukraine nor its strategic depth with large territory and long borders to NATO countries.

Possibly, the Chinese crackdown on the freedom movement in Hong Kong emboldens their resolve. Perhaps, though, it achieves the opposite because almost all people in Hong Kong just seem to continue on with nearly all of their daily activities.

And even active help from the U.S. military may not tip the balance. Chinese missiles and its air and naval assets would most likely dominate the key battleground, which is in the People Republic's extended coastal waters. Without attacking sites on the Chinese mainland, it would take the U.S. way too long to put a significant enough dent into that imbalance in a theater thousands of miles away from home. To see how difficult that is, we can just contemplate the logistics for any nation that would decide to fight the U.S. in the Caribbean.

Third, strategic models agree that, to be successful, any direct military engagement between the U.S. and China threatens to be *highly escalatory.*[153] It likely would cause military attacks on bases near population centers on the Chinese mainland and subsequent Chinese retaliation. This would almost invariably trigger China conducting large-scale warfare, including on the American mainland.

The most often described scenarios assume two main levels of escalation, discounting a third step toward total nuclear war as highly implausible. One of these two escalatory steps would supposedly be a blockade of Chinese supply routes and a localized or regionalized kinetic conflict between mostly traditional military forces of navy ships and fighter planes. The other would be a large escalation resulting in the bombing of positions on the mainland and a Chinese (counter?) attack on Guam, Okinawa, and American carrier groups. The latter may even involve either the Chinese or the U.S. using tactical nuclear weapons.

From there, the vital questions asked are whether the U.S. would trade Los Angeles for Taipei, and whether the Chinese would trade Shanghai for Taipei. The presumed answers usually are "no." However, this jump from localized conflict to global and possibly even total nuclear war ignores crucial facts.

Some of the most devastating technological means of war-fighting available to China and the U.S. involve large-scale infrastructure attacks and drone or missile strikes on military bases on each other's homeland. They would likely result in significant destruction of assets and massive civilian casualty rates across the heartland of either nation. This would not require nuclear weapons. A U.S. blockade of China's naval supply routes, for example, may even make such infrastructure attacks in the U.S. more likely. They would elevate China's most logical counter-option to equivocate and disrupt American supply routes - inside the United States.

Such supply routes and supply chain disruptions have historic precedents in fights among world powers. World War I is one such example.[154] It had far-reaching consequences for militaries and civilians alike.

Differences and Similarities to 1938

Assuming the above three statements are correct, which we will hear more about later in this book, why then does hardly anybody talk about the most likely scenario? This would be that the current trajectory makes a war likely that the U.S. would lose in Asia and that would have massive consequences on the American civilian population?

Instead, most experts, pundits, academics, analysts, and government officials, particularly in the U.S., keep telling almost exactly opposite stories and fairytales. They claim either China backing down or a conflict remaining localized and, in the end, won by the U.S. and its allies. This just sounds like the similar spinning of selective facts in the Europe of 100 years ago.

Why is there little or no talk of the most logical consequences of the facts that most experts agree on, including military ones?[155] Is it really too big of a step to imagine that the unthinkable could become reality? That a Chinese grabbing of Taiwan would be possible, and could even succeed? And that this could turn into a transforming and genuine watershed event for the international system, America's alliances, and our defense posture? Not to speak for our self-image?

When I started researching and writing this book, I had a general understanding of the geopolitical situation, the regional economic and military balance of power, and the economic, technological, and military trajectories of China and the United States. Confirmed through anecdotes and conversations with locals and thinkers from around the world, I developed a tentative hypothesis. The deeper I dug into the details, the clearer the picture got. I became ever more frustrated and concerned about the attitude of those associated with the U.S. Foreign Policy Blob.[156]

Overwhelming evidence kept piling up. And I have to conclude that, for an informed person, the all-too-common defying of reality takes a lot of self-delusion and ignoring of hard facts. Ideology and academic theorizing rationalize an alternate reality. This then warps what a nation's most basic ethical priorities should be: protection of its own citizens and of their freedom.

Here is a simple answer to those claiming the U.S. military's global superiority and moral obligation to fight: the emperor has no clothes! The fight is extremely unlikely to be won. Rather, it will probably result in much more terrible consequences than commonly considered. And so, I must be like the little kid in the famed fable and state the obvious: we cannot and therefore must not directly engage in a fight over Taiwan. Not now or soon. Ideally never.

When they hear statements like the one I just made, many American foreign policy observers and thinkers keep waving terms like "appeasement" and "Munich 1938" to justify hardline positions, military posturing, and even wars. They consider it mandatory that we stand our ground. Their instinctive response is exclaiming an emotional "never again!" But the decision not to fight Hitler in 1938 was *not* appeasement. Even in retrospect, it was a *consequence* of appeasement and a logical choice. Great Britain was in no position to fight Nazi Germany in 1938 successfully, because of decisions it had made earlier in the 1930s and before.

This is a very delicate predicament, because Great Britain was not yet ready to fight one year later either. But then, it was at least somewhat better prepared to confront Germany so close to its own heartlands, aligned with a Japan that threatened most of its Empire. In 1939, Britain felt there was little choice in the matter. In comparison, the United States has in the early 2020s the luxury to be geographically far away from its major rivals, even if America's principal allies are not.

Just because we *dislike* reliving Munich does not mean that we do not already have reached or passed a comparable point. In 1938, Europe's Western powers had spent almost two decades feeling content with their victory in 1918 and the near-total disarmament of Germany up to 1933. A similar strategy of appeasement also was the exact, almost explicit, strategy the United States pursued vis-à-vis China since the 1990s, when Bill Clinton sent a couple of carrier groups through the Taiwan Strait in 1996.

And like Britain in the late 1930s, today we are not in a position either to engage directly in a fight over Taiwan. This time, at least partly, it is because of decisions we made year after year, for the two decades since around 2000. This does not seal Taiwan's fate yet. Similar to China's edge over the U.S. on East Asian battlefields, Taiwan has highly significant asymmetric advantages that disproportionally strengthen its defensive position. Whether this is smart or not, it also can receive American logistical and intelligence support. Such help could even extend to supplies and indirect support comparable to Ukraine during Russia's 2022 attacks.

But for a direct fight between the U.S. and Chinese militaries in the region surrounding Taiwan, our American situation today is even more extreme than Britain's then. Our forces in the theater are far away from our homeland, as I keep mentioning. And although the current trajectory of relative power is reversible, until today it has continued to move to our disadvantage, particularly economically and technologically.

This makes it also more essential to draw the right conclusions. In this context, "right" means "smart," not a self "righteous" easy way that leaves us on autopilot toward a conflict while we cross our fingers and hope for the best.

We may not like this, but it should not surprise us to face such a dilemma. Our society seems to focus energy and money on politics and often secondary and comparably minor issues or policies that do not strengthen our hard power. When our opponents and enemies do the exact opposite, then naturally they will continue to increase their hard power position compared to ours. On both accounts, this is exactly what has happened for at least 20 years between China and the U.S.

To change this, we cannot wish reality away. First, we must clean our own house and rebuild our own economic and technological strengths and defenses. Then, internationally, we must re-establish a strong and widely appealing genuine liberal system worth its name. Such steps must center, above all, on our underlying infrastructure and hard power. All related actions must primarily consider the context of the future, not of the past.

Of course, and also for at least a decade, all these words have been popular political talking points in the United States. But talking points they remained, and mostly still are. Even today, and notwithstanding some recent steps in the right direction,[157] we mostly keep *talking* while continuing to take actions that undermine and contradict what we say. Often for short-term political or commercial gain, we then have no qualm savaging our political opponents and advertising to the world an image of their incompetence or even viciousness. We then justify redirecting resources by denouncing the high standards and effectiveness of our socio-economic system.

It is a truism that we live in bubbles and echo chambers. They frequently create comfortable and often self-serving alternate realities. But outside such cartoon worlds, people *can* get hurt and actions have real consequences. We pay dearly when we ignore time-proven hard truths and empirical facts while elevating lofty secondary cultural goals whose achievement require economic and technological strength first. And in East Asia, we see the results, more consequentially than even in Afghanistan, Iraq, or Vietnam before.

We can and should support Taiwan if it gets attacked. We must show where we stand, help where we can, and clearly show that we consider such an action not only incompatible with our values but also a long-term geopolitical threat to our society. And if we really believe these words, we should act on them, within reason.

But directly fighting a war over Taiwan would be a costly bet, and therefore *foolish*.

If Taiwan loses, and the PRC prevails, then we lost an important battle. At least, though, it would then be one that we hardly could have won militarily, anyway. Losing sucks, but it would be a huge, irresponsible gamble to lose while risking the terrifying consequences of a likely major global war, a World War 3.0. We will neither succeed nor feel better if we passionately throw away the lives of some of our best fighting forces and potentially, or even probably, also of massive numbers of civilians.

Whether or not we like it, the following chapters will show why this is the truth.

9

Military Technology

Military, Technology, and Military Technology

In 1932, the just recently defeated country of Germany kept no meaningful military forces. The Treaty of Versailles still limited the German army to 100,000 lightly armed soldiers, and it officially had no heavy equipment or air forces.[158] Part of Germany was occupied. It just went through hyper-inflation and still was paying sizable sums of reparations to the victors of the Great War of 1914 to 1918. The country was isolated, surrounded by hostile powers.

Merely one decade later, the Third Reich's rebuilt industrial capabilities had established the temporarily most capable and powerful military on the planet. By 1942, it had conquered almost all of continental Europe and attacked Great Britain with air raids. Germany endangered the British Empire's economic lifelines through submarine warfare, military campaigns in North Africa and the Middle East, and an alliance with Japan. Its East Asan ally had conquered most of East and Southeast Asia and threatened the British hold on the Indian subcontinent and Australia.

Germany developed some unmatched military technologies, like jet planes and rockets. It introduced novel war-fighting strategies and started a nuclear program. Until the end of the war, this just recently weak nation's scientific and technological capabilities enabled its military to hold off numerically and economically far superior and mostly better-supplied forces.

Germany did all this within a few years, less than a decade.

Fast-forward 60 years. In 1995 and 1996, the People's Republic of China conducted threatening missile tests in the Strait of Taiwan. But when President Bill Clinton sent two aircraft carrier groups and an amphibious assault ship into the vicinity of Taiwan, China had to back down. The Nimitz, the Belleau Wood, and their support groups sailed through the Taiwan Strait, while carrier Independence stayed close.

This was China's own "Sputnik moment" and galvanized its leaders' resolve to prepare and change the military balance in the region. Unlike the Germans two generations earlier, since then the Chinese PLA and the PLAN had the luxury to spend not just 10 but 25 years during their buildup.

And they used this time wisely. Fueled by economic growth that expanded the Chinese economy 20-fold (compared to America's tripling), the power relationship no longer is the same. In the relevant battle theater, it likely even has reversed. Articles in U.S. military magazines, think tank analyses, assessments by military experts from the U.S. or allies like Australia, and war games simulations keep pointing at this as most probable.[159]

The conventional military balance in the region has flipped and by now strongly disfavors the United States. As uncomfortable, and against the grain, this sounds to the Western public and many policy analysts and politicians, the evidence seems so overwhelming that ignoring it must be seen as irresponsible.

This is highly relevant when trying to make smart foreign policy decisions and prevent a major war in East Asia. According to almost all evidence, we could not win it, now and there. Misunderstanding or misinterpreting the other side's military capabilities can therefore lead to grave policy and strategic mistakes. It can easily trigger a completely preventable blood bath. Or worse, it may expand into each other's heartlands, hitting the Chinese mainland and the continental United States.

There is no inherent reason America and the West could not also make quantum leaps like those achieved by Nazi Germany in the 1930s and the PRC since 2000. But it would require a significant shift in our approaches and priorities. Even then, little we can do today would alter the situation within the next few years. In the meantime, we also will have to consider that technology keeps changing what terms like "weapon" and "military capability" exactly mean.

Are These Still Weapons?

Usually, we express military might as the ability to use kinetic forces that destroy targets and kill humans. Arms kill or hurt through direct physical impact, or indirectly by destroying people's ability to survive. In the commonly used context, this means explosives, bullets, rockets, missiles, and similar weapons.[160] The more enemy infrastructure or assets a nation can physically destroy, and the more people it can hurt and kill, the stronger it is. Millennia of human conflicts unfortunately proved all this.

This is different today, though. In the 2020s, similar or even more far-reaching destructive capabilities no longer need to use conventional military hardware. We can create comparable effects using computer software or "dual-use" commercial systems.[161] Their execution can be digitally choreographed and delivered using communication networks. Exploiting weaknesses of modern complex electronic equipment and networked infrastructure can maximize their impact. This enables a dedicated and capable enemy to effectively take out electricity networks and cut energy supply to military bases, and to consumers and businesses. It also can bring down communication networks, fry computers, and prevent the delivery of goods and products essential for the survival of a country's population.

Even more effective would be combining such new kinds of weapons with other capabilities outside the traditional realm of warfare. For example, these may include sabotage by paramilitary actors that fire rifles at electrical substations or blow up pipelines with remote-controlled drones. And they may also involve semi-autonomous and fully autonomous drones and drone swarms. These can either take off from other continents, deploy from space, or launch inside our country as variations of commercially available and mass-produced products.

In extreme circumstances, the terrifying effects of such attacks can be similar and equivalent to a successful massive bombing campaign, with hardly firing a shot.

This does not take much imagination or difficult-to-acquire specialized knowledge. The most relevant information and data is widely available. It makes it comparable to the massive impact of the Covid-19 pandemic, which happened and surprised so many people. Ironically, coronavirus research had been ongoing and the risk of a pandemic had been predicted, frequently and publicly.

To assess all this, we just must listen to what smart and respectable analysts and thinkers say about this topic, considering their respective areas of expertise. These are not enemies of the West. They include some of the World's foremost scientists, and people with high security clearances and deep knowledge of our defense establishment and military capabilities. Among them are many former or current members of the military itself.

When we then also consider trends and developments of emerging technologies, it quickly becomes clear how one would go about deliberately attacking and destroying key infrastructure and cause harm. The basic attack vectors, the targets and strategies pursued, can be derived from publicly available information. This reveals crucial facts about our electrical grids, oil and gas pipelines, fiberoptic cables, or water supply. And some of the biggest civilian advances have dual-use capabilities. For example, both the military and the civilian sector can benefit from AI, drone technologies and robotics. Both can take advantage of just-in-time supply chains, the billions of devices of the Internet of Things, satellite and space technologies. And both can be attacked in similar ways.

All one needs is pockets that are deep enough, and a sufficient amount of motivation. An advanced nation state that considers itself attacked would have plenty of that. This includes China, although many of such capabilities are also available to Russia, Iran, and even North Korea.

We would then get asymmetric warfare that attacks civilian infrastructure. The disproportionate damage caused by such relatively low-cost and simple new kinds of high-tech weapons can then be equivalent to the effect of familiar weapons on conventional battlefields.

Traditional war would use satellite-based monitoring and AI analysis capabilities to deploy hypersonic or ballistic missiles against strategic targets like aircraft carriers, ships, and military bases. Modern militaries could also deploy semi-autonomous drones and drone swarms that even can bring down fighter planes and strategic bombers or sink submarines.

Far from the main battlefield, as well, these same *kinds* of weapons can attack the same *kind* of infrastructure, revealing the same *kind* of capabilities and asymmetric advantage. In both cases, the centralized design of our systems works against us. Successfully targeting hierarchical networks can be immensely effective.

Some in the U.S. foreign policy community start "getting" it. Henry Kissinger and Eric Schmidt keep repeating their message from "The Age of AI" in interviews and follow-up articles. They point at the impact of artificial intelligence on international politics and particularly the relationship between the U.S. and China.

Although it is an excellent primer on the topic, though, the book's focus on AI discounts the impact of other emerging technologies and of the design of complex systems. Some of those create similar threats (and enable solutions). Examples are IoT, quantum computing, genomics, metaworlds, and blockchains.

Kissinger is credited with designing the visionary opening of China, together with U.S. President Nixon. This is an acceptable description, although also an American-centric interpretation of what happened. A Chinese perspective may describe him as the person in the West who most effectively supported the Chinese leadership's strategy to gain access to Western fertilizers and other critical technology and capabilities.

History validated Kissinger's vision and actions vis-à-vis China. It then is encouraging to see this near centenarian keeping such an open mind about the impact of technological shifts on questions of war and peace.

Military Imbalance: Traditional Weapons

In the medium-term, no later than within the next 10 to 15 years, China is going to reach effective strategic nuclear parity with the United States and Russia.[162] It would gain the ability to execute a full-blown so-called nuclear "second strike," in retaliation for any other power's first strike attack.[163] Even after being hit first, the PRC could then still destroy all primary targets on the other side. These notably include cities, military installations, political centers, and infrastructure and commercial hubs in the United States and probably also in Europe, Japan, and South Korea. China already has a nuclear triad, comprising a fleet of strategic bombers, missiles, and submarines. Turning its current basic nuclear deterrence into effective second-strike capabilities is about scaling only. This is a relatively small step for the biggest manufacturing nation on the planet.

In the Western Pacific, a conventional military equilibrium hardly exists anymore, already today. Nearly all is in China's favor. Most significant is the impact of its hundreds of DF-21 and DF-26 missiles. They limit the U.S. ability to put significant forces into the theater.

Over 2,200 Chinese missiles, including at least 300 advanced medium-range ballistic missiles, are ready to battle U.S. and allied forces in the Western Pacific.[164] They would turn war into an uphill battle and overall ugly affair for America. Even a relatively moderate kill ratio of China's rocket force would cause the destruction of the U.S.' regional bases and of the most significant parts of the American Pacific fleet. Thousands of airmen, sailors, and marines would be dead. Planes destroyed. Ships sunk. Maybe even carriers lost.

And this is just the likely situation today. We also must ask ourselves where the trajectory is going. As just mentioned, China seems to be creating genuine second-strike nuclear capabilities matching or even surpassing the U.S. within the next decade. Its DF-41 missiles can already reach most of the continental United States.

A recent study of the American Physics Society evaluated defensive options against ballistic missile attacks on the U.S. mainland.[165] Although the scenario assumed an attack by North Korea, the authors acknowledged the results would apply to China as well. The bottom-line is that there is no realistic defense against ballistic missile attacks, now or in the next few years.

This does then also apply to such systems that explode nuclear weapons in the atmosphere above the continental U.S. to create an electro-magnetic pulse. There is some discussion about the specific extent of the damage this would cause. At least in theory, though, such an EMP could immediately disable and partially destroy much of the electrical systems within large portions of North America.[166] Since this would include the control systems for almost all critical civilian infrastructure, the negative consequences could be of a near-unimaginably massive scale.

For the likely battleground in the Western Pacific, EMPs seem realistic options also, for both sides.[167] Particularly for China, this could send a powerful signal to the U.S. EMPs can be massive when caused by nuclear weapons. But they can also stay limited and hit smaller targets when delivered via drones or trucks. And, for example, even exploding a nuclear device in the atmosphere above Taiwan may not formally be considered the equivalent of a "regular" nuclear attack. It may not directly cause any casualties, but only fry electronic systems on the island.

Sticking to more traditional weapons systems, the most significant question is whether China's anti-access/area defense (A2/AD) capabilities close to its shores will continue to grow faster than American and allied offensive systems. Unless Western strategies significantly change, this seems likely. During the past couple of decades, military defenses have become more economical than offensive weapons, and overall superior to them. In simple words, and in the just-described context of traditional military weapons fighting each other, offense costs more than defense, much more.

At least in theory, America and its allies would strategically play defense when aiding Taiwan. Tactically and in the larger theater, though, it is the other way around. Overcoming China's home field advantage would require *offensive* strikes that actively counter and prevent attacks launched from the Chinese mainland. And when U.S. aircraft carriers approach the battle zone, China's counteractions would fall into the category of defense.

With this context, a carrier must defeat all attacks by comparably inexpensive weapons, while just one hit could take the ship out, with all its personnel and weaponry. Quickly, such economics no longer add up.

We often count large, centralized platforms like surface vessels by the hundreds. In the same way as our armadas of sophisticated strategic bombers and fighters, these platforms cost up to billions and in their aggregate trillions. On the other side, even thousands or tens of thousands of small missiles, mines, and floating or flying swarms of drones still cost only a fraction of these amounts.

As an extreme thought experiment, even one *million* drones costing $10,000 each would not yet reach *one percent* of the overall costs of the F-35 program.[168] If each drone would cost 100 times as much, *$1 million*, the cost for one million of them would still be below the totals for the F-35 platforms. But the street price of good commercial drones is much lower, in the thousands of dollars. These are not massive, but nimble and flexible. Small and semi-autonomous weapons have therefore become the main means poised to take out the major platforms.

The U.S. also has a hard time keeping up with China's acquisition of more traditional weapon systems. Recent successful tests of hypersonic missiles, in combination with American failures and ended programs, suggest that China may be ahead in the development and manufacturing of such weapons. Meanwhile, new Chinese attack submarines keep coming online faster than the combined total of new submarines that the U.S. and all its allies are building.[169]

China hedges its bets and keeps expanding its traditional weapons as well. In the likely theater, the PLA not only matches but exceeds numbers of the U.S. and its allies, on almost any level. But even globally, its blue-water navy, still mostly "bottled up" in its coastal waters, has more ships than America's. It keeps approaching the U.S.' overall tonnage and capabilities.

New and Asymmetric Weapons

Regarding traditional weapons systems, I will leave it at this: escalation options are manifold. Nuclear missiles are only one of several technologies capable of inflicting crippling attacks on the U.S. mainland and its population. But even if we discount biological and nuclear weapons, fighting China in Asia is daunting. Chinese missiles have longer ranges and are turning hypersonic. The PLA has numerical superiority. China is a peer power that has the home-field advantage where it counts.

But China's ambitions go far beyond the traditional. In the before-quoted chapter 8 of his 2022 book, Kevin Rudd points out that President Xi's guidelines for the PLA as outlined in the 2015 "China's Military Strategy" publication focuses on "forward defense." The fundamental condition would be "informationized warfare." In Rudd's words, this means "long-range, smart, stealthy, unmanned weapons and equipment and [...] cyberspace." The common interpretation of such statements is that they mean the use of such weapons against U.S. and allied ships and planes approaching the Chinese mainland.

This seems illogical to me. Would America consider it *forward*-defense to target ships sailing toward its coast in the Caribbean or 1000 miles from Los Angeles? Rather, "forward" usually means geographically close to enemies or even on the enemy's territory.

Following that logic, we get long-range and covertly launched drones and drone swarms, operating autonomously, supported by space-based weapons, satellite surveillance and control systems, and aided by cyberwar tools, with the goal to asymmetrically maximize damage *inside the American heartland.* And why not? Once war is waged at China's shores, or even inside its borders, and once the PRC's global finances, trade, and energy supplies are curtailed, everything will be on the table, except nuclear and biological weapons.

Hopefully. Let's look at the following with the context of Chinese forward-defense using asymmetric weapons, meaning weapons that are cheaper and more effectively to deploy for attackers than defenses against them.

Bioweapons

Besides nuclear weapons, biological warfare is the other elephant in the room. Increasing, although not fully conclusive, evidence points at Covid-19 having been a designed virus and experiment gone wrong in a combined effort of the global science community to research Coronavirus capabilities. For our analysis here, it is irrelevant whether it happened this way. What matters, though, is that it is a feasible scenario, and technically *definitely possible.*

Using biological weapons has long been a staple of military research of rich countries. Today, though, advanced genomics, big data, and artificial intelligence enable several countries to develop and deploy custom-designed biological weapons after vaccinating one's own population. China is one country with such capabilities, and Russia and the United States are others, and probably not alone in that.

At least in theory, therefore, technology already opens a new range of capabilities. It would require a combination of genomics, artificial intelligence and machine learning optimization algorithms, and massive databases containing DNA and health information. All this is very difficult to do. But it no longer is science fiction and shows how fluid technological advances and military actions can become.

The horrors potentially inflicted could hardly be limited. This is among the most urgent and consequential existential threats that humanity faces at this moment. In that, it rivals nuclear war, the depletion of topsoil, a hit by a large asteroid, and the potential consequences of an effective electromagnetic attack. In Wuhan, China has one of the three most advanced biological research labs of the world. It has more engineers working in this field than any other country safe the United States. [170] Its centralized and totalitarian political system makes contact tracing, aggressive containment, vaccine mandates, and their execution easy.

All this is just the beginning. Great power rivalry, and therefore also a global conflict, may not even remain confined to our planet.

Space

For civilian and military uses, the rules and tools change when we leave Earth. Space technologies have applications for communications, observation and surveillance, and science. They also create the foundation for ever more futuristic commercial technologies, from energy generation to resource exploration. All this is important for matters of the military as well, and ultimately for warfare.

That being so, China has become a significant space power. Japan and India have significant space programs also, as have European countries, the UK, Israel, and others. But China is catching up and in the process of passing the traditional space power of Russia. The PRC is about to complete its first multi-module space station and has independent space launch capabilities with its Long March missiles. China is even establishing what seems to be a permanent base on the moon, or near it, in a cislunar orbit. [171] Among its ambitions is to explore energy production in space, whether through space-based solar power or by mining Helium-3 on the moon as a raw material for nuclear fusion reactors.[172]

Space technologies have dual-use applications, for civilian and military purposes. Geopolitical analyst Brandon Weichert describes many of these mutual "space war" capabilities in "Winning Space: How America Remains a Superpower." [173] For example, the Chinese military just put in service its own "Beidou" GPS-equivalent satellite-based navigation system. [174] It also has successfully tested anti-satellite weapons that can blind U.S. forces and interfere with its communications. At least in theory, space enables China to deploy EMP weapons in the Western Pacific and even over the U.S.

Even if America could strike back similarly, on the Western Pacific theater, this would still be a net advantage for China. The U.S. would fight far from home and even from its own bases. This leaves it more heavily dependent on space-based communications. China could leverage its home field advantage and fall back more easily on ground-based systems. Some of the American military's key technological advantages depend on well-functioning fully networked communications and computing systems supported by satellite surveillance.

China uses similar systems, which also makes it vulnerable. But today's technologies still ensure that communication on the ground can often be easier than via space or air. Large private constellations of thousands of Low-Earth Orbit (LEO) satellites, like SpaceX' Starlink, are not yet large or reliable enough. In the likely Western Pacific theater, all this works in the PRC's favors. It makes it unlikely that any foreign power, including the U.S., could gain air superiority over large parts of the Chinese mainland. This therefore protects the PLA's command-and-control infrastructure, and many of the launch pads and bases for its missile and air force.

Such technologies can give China an advantage. Even with both sides' space communication wiped out, the PRC would likely still be able to coordinate defense communication. It could use a combination of fiberoptic cables, cellular towers, and 5G (and soon 6G) mesh communication webs.

Barring a major escalatory step, the U.S. and its allies could at "best" hope to take out or block local satellite-based communications and command capabilities, or parts of the PRC's overall space-based systems. What would wipe out many capabilities of U.S. forces in the Western Pacific would be closer to a major inconvenience for China's PLA.

All this means that, at least theoretically, China can severely disrupt American military surveillance and communications, particularly those relevant to the likely theater of war. The U.S., though, cannot take out all Chinese equivalent infrastructure in the region. That the battlefield would be close to China provides the PRC with an enormous advantage.

To achieve parity, we are back to the ultimate escalation step: the U.S. would need to strike at China's homeland. This would involve offensive air and missile strikes, sabotage, drone and cyber-attacks deep into Chinese systems on the mainland, including its new mesh-systems and fiberoptic network. The latter would be similar to what the Chinese could do to America, or at least what it could try. But the task would be easier for China because the U.S. systems may traditionally have been more capable, but nowadays are older and more centralized. This makes them more vulnerable. In the 21st century, decentralization beats centralization.

Just ask the people of Afghanistan today.

Sabotage and more

We can go one step further. China has most likely many agents operating in the U.S. right now. This may be virtually and in cyberspace, or physically on the ground. We do not need to stretch our imagination to see this allowing the PRC to disrupt a critical number of key nodes of the centralized American power grid, its fossil energy infrastructure, and the Internet and communications network. The most dramatic effect could be the complete destruction of domestic logistical supply chains.

These systems are not built to be secure. Rather, they are optimized to reduce cash flow needs and deliver products just-in-time. Among them are consumer products like those ordered on Amazon. Others include food, energy, raw materials, and intermediate products for businesses. And a significant portion of the Internet's underlying infrastructure is *built to maximize - advertising revenues for Google and Facebook.*

Additionally, and equally important, Google's and Facebook generate most of their revenue by *collecting* and *selling* data. This revenue model requires the opposite of privacy and therefore also security, since those two terms are interdependent. Enemies could either try to hack the paragons of American technological prowess, or they could instead *buy* much of the relevant information from them.

For sabotage, the number of basic attack vectors, or scenarios, is tremendous. They include attacks via sabotage, drones, missiles, cyber and computing tools, electromagnetic denial, anti-satellite weapons, and EMP.

Hurricane Sandy caused havoc on the U.S. East Coast, and so did the hack on the Colonial pipeline in May of 2021. What would happen if several or all major pipelines were down? If strategic attacks on our grid would be added on top of it? In California in recent years, we had widespread outages due to "attacks" from - trees. Branches of trees fell on power lines and then caused large fires that hurt and killed people and caused economic damage. All it took was not to cut down trees next to power lines, because of a combination of environmental protections and misplaced corporate priorities.

We have over 2.7 million miles of power lines and 450,000 miles of high-voltage power lines in the U.S. and Canada.[175] They all are connected in a far too hierarchically structured and therefore vulnerable grid system. How could we secure them all?

More about these topics in the chapters 12 and 14.

The combined effect of the above shows a stunning range of direct and indirect novel military capabilities and options for a nation-state combatant like China. These cover the full spectrum. They range from missiles, satellites, drones, fighter planes, submarines, surface ships, to a nuclear triad smaller but in some ways more modern than ours. In most of these fields, Chinese numbers are on a trajectory to approach and exceed our own within roughly a decade from today, wherever they have not already done so. China has asymmetric advantages plus the ability to reduce or even destroy U.S. strengths in the war theater.

The PRC can hit the American homeland through sabotage, cyberattacks, and destruction of critical infrastructure. Although it also can hit with its nuclear arsenal, nukes are not even the most potent of its weapons. If, against all odds, America and its allies establish a successful blockade, sink most Chinese ships, or take out bases on the mainland, the PRC may still prevail. It would have outstanding abilities to match U.S. moves and militarily retaliate in kind, or attack in ways that have an equivalent effect.

It then does not even matter whether we win battles, and Taiwan remains free. Our homeland would be under attack.

War Games and Chinese Movies

Considering the above, it should not be a surprise that most military strategist stress that almost all war-games show the U.S. losing a direct confrontation with China in East or Southeast Asia. In Christian Brose's "Kill Chain," he is even more explicit. Brose talks about "a nearly perfect record: we have lost almost every single time."

Losing war-games does, of course, not yet mean that China would "win" in any traditional sense. And there is a bit of grumbling that some of these war-game scenarios are deliberately skewed to make it impossible for the blue team to win. But simple arithmetic shows that the war games' outcomes likely are realistic.

The best we can hope for may be a repetition of the Korean War. But the first part only, the Korean War taught in Western schools and talked about in U.S. foreign policy circles. In this, American-dominated forces prevented almost certain defeat of the South and pushed back the North Korean aggressors after the genius military landing of McArthur's troops in Incheon.

The second part was not as rosy for the American side. The U.S.-led troops did not show restraint. Rather, they crossed the 38th parallel and marched up north and toward the Chinese border. Some talked of crossing the Yalu River and heading toward Beijing. And so, the PRC intervened and turned it instead into an American Long March, southward and in retreat. It then ended when the front reached the 38th parallel again whose crossing by North Korean forces had started the whole conflict.

Those interested in the Chinese popular culture of 2021 may consider watching "The Battle at Lake Changjin," which was the most popular movie in Chinese theaters during October 2021.[176] Yes, the Chinese really won this battle, and the American forces had major casualties. In Chinese propaganda, the Korean War therefore has a different name than in the U.S.: "War to Resist American Aggression and Aid Korea."

The story of the 1950s Korean War suggests that it may not be a smart idea to discount Chinese military prowess near their homeland. But this is exactly what it sounds like when some in the American foreign policy and military establishment keep pointing out, correctly, that China has not really waged any war since 1979 (against Vietnam) while the U.S. military is battle-hardened. They point at China not having experience in managing large-scale military supply chains, nor in effectively moving massive battle groups comparable to American aircraft carrier groups.

The counterpoints should be obvious: The U.S. has not really waged a war against a peer power since 1945, and not against a major power since 1953. While Chinese military supply chains are not just on their own continent but even almost completely inside their homeland (!), American supplies must cross the largest body of water on our planet. This exposes them to enemy satellite surveillance, submarine activity, and remote attacks by swimming or flying mines and drones, and missiles. And all this happens in a geo-strategic context that strongly favors defensive actions over offensive ones.

Although America knows how to coordinate complex movements and actions of naval battle groups, these battle groups are unlikely to get near enough to the Chinese coast to engage. Despite America's F-22 probably being superior to China's J-20,[177] it is doubtful that many of these fighters can get close to the battlefield unless America attacks the Chinese mainland first to take out the PLA's missiles. And then the question is whether we are back to a situation comparable to the battle at Lake Changjin.

We can even get a bit more specific.

America relies heavily on modern versions of generation-old strategies and types of platforms, while having fallen behind in some key aspects of modern weaponry. This even applies to sophisticated weapons and, most noticeably, to the ability to conduct asymmetric warfare.

For example, the American military has recently started using commercial satellite and communication systems as alternatives. This is, in principle, a smart move. However, the largest parts of our military communication and control systems still depend on a relatively small group of exposed military satellite "constellations" that hardly deserve to be labelled that way because they only consist of small numbers of satellites. Similarly, most of our key national infrastructure is centralized and too easy to disrupt. The most notable part is the electrical grid. These risks posed by big and centralized platforms extend even more so to critical infrastructure in the American homeland. All this opens the door to Chinese asymmetric warfare by inviting the PLA to take out a few centralized nodes of critical networks to let the U.S. military fight blind and deaf, and to deliver chaos, destruction, and despair to civilians.

Neither the U.S. government's white papers nor its $700 billion defense spending contradicts this. Or what novel weapon systems being able to survive missile saturation attacks have been developed, put in service, acquired in large enough numbers, and deployed in the Western Pacific? To what degree have we hardened our military bases, let alone our civilian infrastructure? What semi-autonomous drone swarms are deployed or deployable in the Western Pacific? Maybe there are some miracle weapons hidden somewhere. We also hear about tests of some of the new tech I describe in other chapters, like air mines or drone swarms. Though, it rather looks like we have been too comfortable, distracted, and overly bureaucratic for decades to notice how far the world has shifted.

It means little that our fighters are superior in battle, our aircraft carriers better, and our submarines more capable. Most of this may unfortunately be much less relevant when it matters. It often seems to assume the enemy would invite such weapons to a fight and execute generation-old strategies and tactical maneuvers. More likely, the opposite is the case.

China has built missiles that can deny our large naval platforms access to the Western Pacific. It can conduct cyberattacks and use their submarines to intercept our supply lines in the heart of the Pacific and off our own coasts. The PLA's anti-satellite weapons can blind the U.S. on the battlefield and take out critical communications. The Chinese can also intercept or take out our civilian communications and threaten supply lines and critical infrastructure on the North American continent itself. In a world dominated by digital and space technologies, the seas no longer protect us. Relative to other means of warfare, navies do by far not carry the same punch as they did in the 1940s. Our society is more dependent on space and the digital commons than on the blue oceans.

America sure could inflict horrible damage on Chinese forces. But in the large picture, all we can hope for is that China bungles this up and that we shoot down enough of their missiles to just lose part of our own fleets. Or that some tactics or weapons technology will work more effectively on our side rather than theirs.

Above all, there is one reason the U.S. keeps losing in simulations: it does not have enough forces in the likely theater and therefore always runs out of *time*. Time is short in five areas across a range of important variables.

- First, it will take weeks even for our nearby fighting forces to arrive, comprising Navy, Air Force, and Marines in Western Pacific bases. The estimate for our Japan-based carrier group is about 17 days.
- Second, it will take several months to move the required additional platforms and massive numbers of equipment and personnel across the largest body of water on the planet to sustain the fight or even attempt building regional superiority.
- Third, it will take many months to several years to develop and modernize such weapons systems that can break through and defeat Chinese lines around Taiwan and adjacent islands. When the fighting breaks out, it is too late to develop such weapons.[178]

- Fourth, it will take years to harden infrastructure on the American heartland and make it resilient to a Chinese attack. In the absence of such resilience, much if not most of the war-fighting effort would have to focus on homeland survival. It would re-establish communications lines and digital systems and redirect supply chains for food and energy.
- And finally, building credibility and support in the international community is time-consuming and requires demonstrable success in matters that directly affect people living in other societies. Unlike in the past, when soft power seemed to be a near-unsurmountable U.S. advantage, today the field is much more even than the West likes to believe. China's recent economic success story, the heavy-handed U.S. response to 9/11, and the globally broadcasted hate-fest of American politics has put a dent in much of this. Beyond a narrow group of allies, America cannot take for granted support by "the global community." It is particularly questionable for countries economically dependent on Chinese markets and its economy's continuing growth. It seems premature to assume that "world opinion" would automatically side with the United States. Southeast Asians, Africans, Latin Americans, and eventually even Europeans may back off rather quickly. This is not 9/11, which triggered near-unified goodwill and global support immediately following the attacks on America.

With these five technology-centered points, it is easy to see the U.S. finding itself confronted with rapidly worsening hard facts on the ground. In the meantime, it has inadequate bases, insufficient and in some places inferior weapons systems, and follows outdated military strategies. Easy to compromise communications and control systems leave it with few options at a time when social media and algorithm-optimized and directed international support can erode faster than America can move aircraft carrier battle groups across the Pacific.

And about that, carriers approaching China, let's now see how that would work out when the shooting starts.

10

Battlefield

Defense Rules - and Asymmetric Weapons

A single missile streaks into the sky from its launch pad near the Chinese coast.

It is not the only one. Far from it. Many others hit targets on and around Taiwan. But this one is different. Like many others, it heads west. Although not directly west, but more to the north. And it flies higher, faster, and farther, aiming at the open Pacific Ocean.

If it had eyes, it could see the island of Formosa, Taiwan's main island. There, most of the fighting is going on since the People's Republic of China started its invasion. But the missile is not concerned with what is happening there. It targets what is approaching the battlefield.

The USS Ronald Reagan (CVN 76) is a nuclear-powered Nimitz-class aircraft carrier with a home port in the Bay of Tokyo in Yokosuka, Japan. It is part of the United States Seventh Fleet and the flagship of Carrier Strike Group Five (CSG 5). The mighty ship displaces slightly above 100,000 long tons, carries up to 90 aircraft and helicopters, and is home to over 6,000 personnel.

Nimitz-class carriers are floating military bases of near-unrivaled capabilities. Together with a global network of about 800 ground bases, aircraft carriers are America's primary tools to project and apply military force far from its shores. The U.S. Navy says specifically about CSG 5's role that its platforms "have a higher operational tempo and are an average of 17 steaming days closer to locations in Asia than naval forces based in the continental United States."[179] Sometimes, the U.S. has one or two more aircraft carriers in the Western Pacific. If not, the USS Ronald Reagan and CSG 5's other ten surface combatant ships are therefore the likely first large naval force that can directly confront the People's Liberation Army Navy in case of a war. This is America's only forward-based carrier, without a home base on U.S. soil.

And so, when Taiwan were to get attacked following the scenario outlined above, CSG 5 would move. Despite China's warnings, the carrier and its flotilla would prepare for combat and speed toward the battleground around Taiwan. According to dominating doctrine, no matter what, America is going to show flag on the battlefield and show its willingness to confront Chinese forces directly.

But China will have none of this. Its satellites and drones will have followed CSG 5 during every step. They have given clear target coordinates, which they continue to refine until the last moment to account for the fleet's movements. And so, long before the USS Ronald Reagan reaches Taiwanese waters, it has to confront its most formidable enemy: the Dong Feng 21 (DF-21) heading its way. The DF-21 missile is a ballistic weapon, meaning it leaves the atmosphere when fired. This highly parabolic trajectory limits the time for advanced warning, and its hypersonic speed of up to Mach 10 and its maneuverability make it very "tricky" if not impossible to defend against.[180]

Less than 15 minutes after its launch, the missile, whose launch I described above, is therefore likely going to encounter the battle group. If it reaches the USS Ronald Reagan, the consequences will be devastating.[181] A direct hit of the carrier's main deck may not even require an explosive warhead.

Some calculations suggest that the missile's immense speed and kinetic energy coud keep it going all the way, punching a hole from the carrier's deck to the bottom of its hull. It would almost immediately incapacitate the most massive ship in America's arsenal. And the Ronald Reagan has few defensive options. Its most promising ones may be to take out, or confuse, sensors and systems feeding the incoming missile's targeting computers. These are essential to give such a weapon a chance to hit the moving target of this naval vessel.

The above is far from detailed enough to describe the specifics of what would happen in such a battle. Likely, the U.S. will attempt to deploy anti-satellite weapons, use decoys and air defenses, jam the incoming missile's electronics, and interfere with the Dong Feng's guidance and targeting systems. Aegis missile defenses and other weapons will try to shoot down the missile and may even succeed.

But the underlying capabilities of a DF-21 are as just described. That part definitely *is* realistic. More likely, even, it will carry a conventional or nuclear warhead and may therefore not have to hit as precisely as I just outlined. And such a Chinese missile attack can succeed even if countermeasures work and a single DF-21 would fail for any reason.

Even technical shortcomings, or the U.S. Navy's defensive measures, resulting in the DF-21 failing may not matter much. The Pentagon assumes the PLA has over 300 DF-21 in its missile arsenal of over 2,000. If one misses, there are plenty others to follow up with. And among these are even more capable alternatives, like hundreds of DF-26, or hypersonic gliders like the DF-17. Against such speed, power, and numbers, we must assume that the U.S. has neither sufficiently competent defenses nor matching capabilities in the theater. Partly because of the Intermediate-Range Nuclear Forces (INF) treaty that President Trump eventually cancelled, America did not even *develop* equivalent weapon systems, let alone station them.

Most important for the above scenario, China does not need to use nukes. Not even a warhead may be required, as depicted in the above scenario. The PLA has several options on how to sink an aircraft carrier.

This capability gave the DF-21 its nickname.

Carrier-killer.

The Arsenals of War and Escalation

"Fighting for Taiwan" can mean many things. On the one extreme, it may mean dropping nuclear bombs at each other's population centers. As the most drastic statements describe it, this would "trade" Los Angeles for Taipei, or Beijing for Taipei. The other end of the spectrum may not even involve kinetic weapons or actions that *directly* kill human lives.

On the lowest rung of the escalation ladder are actions formally not even considered acts of war. These would be the freezing of financial assets, trade sanctions, and the closing of ports for each other's trade. Once we are going beyond these, though, we are quickly approaching direct warfare.

There are a range of potential "non-kinetic" options. The U.S. can share intelligence with Taiwan and supply weapons and supplies.[182] This is the grey zone. America may further conduct cyberattacks on Chinese military and civilian infrastructure, establish blockades, cut fiber optic cables and destroy communication satellites. Both sides may try to blind each other's navigation and electronic systems either on the battlefield or at a larger scale using electro-magnetic pulses. Some of these may already tackle China's civilian infrastructure on the mainland, which is a battleground I will explore more in the next chapters. By now, we are in the middle of all but "real" warfare. For most what I described so far, it may not have taken us longer than several hours to reach that stage.

Next on the list, already, are then direct engagements between U.S. and allied military forces on the one side and the PLA on the other. The U.S. may deposit mines. Its submarines and anti-ship missiles are likely to sink Chinese ships. American planes may shoot down many of its Chinese counterparts.

Even the use of tactical nuclear weapons on either side seems possible. Many war-game scenarios consider China using tactical nukes to strike at U.S. carrier groups or to hit American bases in the region. Frequently, this assumes the PLA trying to prevent an impending defeat by allied forces as carrier groups approach the battlefield. But as the story I began this chapter with has outlined, the PRC has missile systems in the region that at least on paper seem capable of stopping the carriers without the need to involve nuclear weapons.

The U.S., though, does not have directly comparable options. One main reason is the just-mentioned INF treaty between the U.S. and the Soviet Union (and later Russia). It prohibited the development of such weapons systems until the U.S. canceled it in 2019.[183] Therefore, a realistic alternative scenario is a first use of tactical nukes by the U.S. military. This would compensate for a more likely regional battlefield advantage of China. But it would also mean using nuclear weapons in Chinese waters or even on the Chinese mainland. This would be in close vicinity to China's major population, industrial, and commercial centers. It would immediately raise the stakes to the highest level possible, because from a Chinese perspective such an assault is nearly indistinguishable from a strategic nuclear attack. A nuclear strike at Chinese forces could therefore trigger a strategic response using China's nuclear triad on the American mainland, or the use of biological or chemical weapons.

Nuclear war would always just be half a step away, for several reasons.[184]

The decisive question is whether low-end non-military measures could be effective without escalation that targets U.S. and Chinese military bases and platforms. When would either side simply give up and withdraw from the battlefield, or at least be ready for a ceasefire?

Or would both keep on fighting and escalating like the Western powers did in World War I? There, they mostly just continued with the strategies and tactics they started out with on the man-eater in the southern Alsace, until their countries were exhausted. Almost all options sound unpleasant and unlikely.

Maybe we must stop escalation at the very beginning. Let us therefore look at where and how a regional war would likely start.

The Hotspots: Causes of War

Most pundits see disputes surrounding three geographical areas as the most likely origins for military conflict and subsequent escalation between the PRC and the U.S. Going from north to south, they are the Ryukyu and Senkaku Islands in the East China Sea, Taiwan, and the South China Sea. Brandon Taylor of the Australian National University adds the Korean Peninsula to it and calls these "The Four Flashpoints."[185] Korea is a special case, though, because it does not involve Chinese territorial claims.

The other three are related and impact each other. The core geographical issue, however, is Taiwan. It is centrally located and has a large and advanced economy. Taiwan also has a medium-sized population that, since the 1990s and until today, has no desire to be governed by anybody else and, most importantly, not by another political system.

Compared to Taiwan, other islands mostly just are the sideshow. Grabbing an island or two in the Senkakus or even all Spratly Islands in the South China Sea would merely be a tactical move.[186] It cannot sufficiently address or resolve the larger geo-strategic issue.[187] If it is at all possible, China's ability to significantly improve its strategic position stands or falls with Taiwan. As long as China considers traditional unrestrained blue-ocean access as crucial for its strategic independence, it will therefore pursue a takeover of Taiwan, almost at any cost.

The already mentioned geopolitical, historical, and cultural arguments also open options for diplomacy (see the Postscript "America's Choice" of this book). But China uses also more narrow political ones. The People's Republic of China argues that, since the 1970s, the world acknowledges Beijing as the sole representative of all of China (including, since 1979, the U.S.).[188] It received the Chinese seat on the U.N. Security Council previously held by Taiwan. *Both* sides had then still claimed to represent one common nation that combined mainland China and Taiwan. Today, about 180 nations have established official diplomatic relations with mainland China instead of Taiwan. The United States of America is one of them. Summarized by the phrase "one country, two systems," this moves Taiwan formally, according to a reasonable interpretation of international conventions and laws, into a status of being a part of China.

Obviously, this is not the only way of interpreting the situation. Every one of these 180 nations understands Taiwan's disputed status. In reality, though, their dropping formal diplomatic relations with Taiwan clearly indicates where they are standing on that matter. Most would not oppose unification of Taiwan with the PRC and be unlikely to take strong measures in case of this happening by force.

Apart from Taiwanese resistance, there is just one major impediment blocking China's enforcement of its version of the big picture and asserting direct control over the ROC. This is the U.S. policy of deliberate "strategic ambiguity" and the related active deployment of the U.S. military in the region.

What does that term mean? The answer is simple: "who knows?" This is where the "ambiguity" part comes in. But then American politicians and high-ranking generals and admirals talk about war and actively defending Taiwan. *That,* then, sounds like Sarajevo in 1914. And among these politicians is Joe Biden, the country's president.

Japan has recently ensured its support for the United States in that matter, and therefore for Taiwan's status quo. As Japan's Prime Minister acknowledged on 13 October 2021, Japan considers a potential conflict over Taiwan as a direct threat to Japanese security. He expressed Japan's readiness to militarily defend the island in case of a Chinese invasion. [189] This is in line with sentiments frankly expressed during an interview I recently conducted with a Japanese think tank. It seems ever more likely that Japan would consider an attack on Taiwan equivalent to an attack on the Senkakus and retaliate with full force.

Another position could be difficult to maintain for Japan. The major U.S. base that could immediately engage Chinese forces is in Japan (Kadena Air Base on Okinawa). And the home port of America's only forward-deployed aircraft carrier is in Tokyo Bay.

For the American government, fighting for Taiwan is also legally straightforward to defend. The Taiwan Relations Act of 1979 requires the U.S. to assist Taiwan in its defense against an attack. This includes an attack from China, and all but obligates the U.S. to militarily intervene.

Triggers for War

A shooting war between the United States and China over Taiwan would therefore be difficult to prevent if China would attack Taipei. And a range of specific events could trigger such an attack. The escalation of minor confrontations is certainly possible but would more likely catch both sides unprepared. Particularly, China would not want to be caught that way. The success of a major war would, to a large degree, depend on its ability to establish favorable facts on Taiwan's main island of Formosa as soon as possible.

In theory, a Chinese-Filipino or Chinese-Japanese conflict over commercial activity and rights of fishing boats could spin out of control to the north or south of Taiwan. Also, freedom of navigation operations (FONOPs) in the South China Sea may cause lives lost. But neither involves core interests of any participant. With damage and numbers of casualties in such conflicts likely remaining limited, we can expect them to stay contained. The most probable scenario sparking a conflict would therefore be domestic developments in Taiwan or in China.

Two Taiwanese actions could trigger an attack by the PRC. One is a declaration of independence, which long-standing Chinese policy considers a reason to go to war. The other one would probably be the nuclear armament of Taiwan. Although it has three nuclear reactors and certainly also the money and the skills to do so,[190] though, Taipei shut down for good its nuclear weapons program in 1988.[191]

As of mid-2022, it seems unlikely that either of these two Taiwanese actions would happen. The power imbalance simply is too vast. And most of its people are content with the formal status quo.

This leaves the main potential triggers with China. And there, it would require not much more than making good on repeated and consistent Chinese statements. These say that unification, if necessary by force, is on the agenda and will continue to be pursued. It then seems logical to assume that eventually the PRC would order its PLA to forcefully take over Taiwan.

Continued trends of economic and military power could make this ever more probable. If the most recent trajectory continues similarly to how it has for several decades, China's position would continue to improve. It would keep getting stronger absolutely, and relatively when compared to the U.S. By the early 2030s, China would pass the U.S. GDP even in nominal terms. It would then have even more solidly expanded its lead as the largest military power in the region by almost any standard. At some point, it may then just decide to create cold hard facts and formally establish control over Taiwan.

Unfortunately, the opposite economic direction would likely not introduce a more stable situation either. If China's growth were to slow down, and a domestic crisis was about to break out, this could destabilize its society. Historically, Chinese populations don't take it lightly if their economic likelihood deteriorates. Even just a significant slowdown could therefore challenge the country's traditional success formula and, subsequently, the government's legitimacy. The danger of war may then become even more imminent. A Malvinas/Falkland Islands scenario would be possible, involving a besieged government launching a popular military adventure to close ranks within its society.

This makes the most probable situation a well-prepared but sudden move by overwhelming Chinese forces. They would execute a carefully planned series of military actions that establish the PRC's presence in Taiwan. In the absence of any other credible threat to China, its PLA could clearly consider itself ready to ensure military success. This also includes the repelling of any outside help for Taiwanese forces.

To figure out how this is likely to pan out, we must look at the superpower's options and actions. The following does not replay the many variations of scenarios. It is solely to outline the basic military situation and potential actions during the first days or weeks of a fight.

Setting the Stage in Taiwan and Its Neighborhood

From the beginning, China is unlikely to leave the U.S. with many good options. If the PLA would conduct a successful flash attack on Taiwan, this would establish facts on the ground quickly and make China's job of defense against allied forces much easier.

Chinese preparations for such an action may not be noticeable to the outside world, maybe not even until *after* hostilities have started. Already today, the PLA constantly harasses Taiwanese forces and, at times, has dozens of fighter planes entering Taiwan's air space. Preparing for a hot phase of the conflict, Chinese action could therefore be just a continuation of time-proven Chinese strategies when encountering enemy forces.

The PRC would use its locally overwhelming size advantage and numbers while pursuing "death by a thousand cuts." Not a single action would be substantially more severe than the preceding ones. It would always insinuate that any counteraction by the other side would be an overreaction. And the advantage of appeasingly working with the Middle Kingdom would be immensely appealing.

This is close to what the PRC has been doing since at least the beginning of the century. It pulled many of the most significant aspects affecting the Taiwanese population's daily lives under the Chinese economic umbrella and system. By successfully entangling and effectively integrating substantial aspects of Taiwanese society into its sphere of influence, it established economic facts on the ground and created dependencies. Without firing bullets or exploding bombs, this increasingly blurs the impact of what the most significant and final step would mean for ordinary Taiwanese: political control by the CCP.

Already today, a large part of this has happened. Around 2020, China had solidly established itself as Taiwan's biggest trading and investment partner. Many people have heard of Foxconn, the manufacturer (among others) of Apple's iPhones. This is a

Taiwanese company, operating its main manufacturing facilities on the Chinese mainland. It is a prime example of the PRC's success in co-opting an adversarial society's priorities. Already around 750 years ago, the Mongol conquerors under Kublai Khan succumbed to the same treatment. We should not be surprised if much smaller societies like Taiwan may not be more successful today.

The surreptitious way this works is by aligning economic incentives and daily interactions to a degree that loyalties cross national borders. Political questions then become diffuse. The economies and their supply chains grow even more integrated, and cooperative activities continue to generate wealth on both sides of the Taiwan Strait. A practical question then pops up: what part of one's life really is being limited, or even just touched, if the people sitting at immigration desks switch their uniforms? This is how it worked in Hong Kong for many years. Nothing of substance changed until recently when, of course, China pulled the curtain. But even then, of the seven times 24 hours of a typical week, what portion has noticeably changed? This is not my argument but simulates a thought process that may be going on in Taiwanese minds.

The Fight

Should China use serious military actions, it would likely be via a surprise attack. Shock troops would clandestinely enter the country, using commercial vessels. Additional forces would be on the way using civilian means of transportation, making it difficult to distinguish between civilians and the military.

Possibly, some variation of what the U.S. Congressional EMP Commission describes as "combined-arms cyber warfare" (or also "Blackout Warfare") would assist these activities. Cyberattacks, sabotage, and weapons generating non-nuclear electromagnetic pulses (NNEMP) would take out a large portion of Taiwan's core infrastructure. This would be both an effective tactical military move and a strategic warning to the United States and Japan. It says, "look what we can do to a country that opposes us."

By the time military airplanes, drones, and more PLA operators hit Taiwan, the stage has been set and PRC's ships are already approaching as well.

If things were to go perfectly for China, it would knock out a sufficient part of Taiwanese defenses almost immediately. Chinese troops would land on some of the few suitable beach-heads, or emerge from civilian shipping containers, and rapidly establish control over key infrastructure on the island. If it works well, this would prevent widespread bloodshed.

Otherwise, things would get ugly. In many areas, they could descend into urban warfare. By the time traditional military action would begin, with fighter planes, submarines, mines, and missiles, the tables already may have turned. Now, much of the Taiwanese infrastructure would be offline or occupied by PLA personnel. This also reduces or eliminates potential targets for Taiwanese missiles and fighters that may be in the air or ready to launch. What could they hit? Their own substations, power generators, ports, and airfields?

Of course, nothing ever goes fully according to plan. Just in the first half of 2022, the Russians learned this the hard way during their attack on the Ukraine. Likewise, so did the Ethiopian forces at the same time in Tigris. For the sake of this argument, we can therefore assume that fighting ensues, with waves of PLA planes crossing into Taiwanese airspace, and landing boats approaching mostly defended beaches and their minefields. Missile barrages from the mainland would attempt to overcome defenses, and waves of aerial attacks would rapidly wear down Taiwanese forces.

The NNEMPs, sabotage, drone and cyber-attacks would result in a widespread electricity and communication outage. They would also limit the ability of Taiwanese reserves to join the fight. But Taiwanese forces may hang on, take down fighter planes and sink invading ships to buy at least hours of time. The military of the Republic of China (ROC) may bring back enough power, hold defensive positions, repel operators already on the island, and deploy counterstrike cyber capabilities. But no matter what, sheer numbers will mean that the chances of succeeding will rapidly shrink, unless they get help from the outside. And this means above all the United States and Japan as active participants, likely aided by Australia and the U.K. because of the new AUKUS pact.

Unfortunately, even their combined options are limited, at least in the short term. Using traditional weapons systems, the U.S. military can project force in the Western Pacific through three major ways: bases, carrier fleets, and submarines. Particularly, the base on Okinawa contains immediately available firepower that could reach the main battlefield. Guam and other bases in Japan and in South Korea could contribute as well. Other alternatives also exist, in theory. One example is Agile Combat Employment

(ACE). [192] Among its most important components is that it multiplies the number of air bases, pre-planned and ad hoc. This increases the numbers of locations an enemy would have to deal with. While being promising, though, it is also more complex and difficult to support logistically.

Okinawa is about 400 miles from Taiwan. The Air Force can fly sorties from there. It is too far to enable fighters sticking around on the battlefield, though. The island is Japanese territory. And per the latest statements of Japan's defense minister, Japan would not wait for the PLA to preventively attack Okinawa but interpret an attack on Taiwan as a casus belli and immediately be ready to join the fight.

Guam, however, is too far out for F-35 fighters to reach Taiwan without refueling outbound and during the return. Since the tankers do not have their own defenses,[193] this would be very tricky and likely impossible to achieve. Following the traditional playbook of America's recent wars, it would therefore mainly stay designated as the staging ground as the U.S. pushes additional forces into the region.

There are three considerable problems with counterattacks from either Japan or from Guam. The bases themselves are vulnerable, they are far from continental U.S. supply lines and reinforcements, and the U.S. has little time. Okinawa and even Guam are within range of highly potent Chinese weapons that were specifically designed and built to hit and destroy them. Against some of these, particularly hypersonic and ballistic missiles, the U.S. has no sufficiently functioning defenses.

Time may work out if China would only slowly increase its pressure on Taiwan. Even then, though, this would depend on China failing with destroying Guam. But this is exactly one objective the Chinese military has been working toward for decades. In the above scenario, the U.S. simply would not have the luxury of slowly crossing the Pacific Ocean and amassing forces in the theater, unimpeded by Chinese missiles and attack submarines. The United States would be unlikely to get the weeks and months it takes to move the required large numbers of military assets close enough to make a difference.

The carrier-killer missile described at the beginning of this chapter sums up the situation. And it is neither a new concept nor a secret.

While I studied in the Masters Program in International Relations at the University of Delaware, in 1990, the program's head and naval expert James K. Oliver talked about the vulnerability of aircraft carriers, "also known as targets." This was over 30 years ago. Australia's Hugh White echoes this sentiment and says in more stark words that, "I don't think it is probable that the boys in Pearl Harbor would even risk" their carriers. As professor at the Strategic and Defense Studies Centre of Australian National University, he was the principal writer of Australia's 2000 Defense White Paper. White assumes that U.S. Pacific Command would try to keep their carriers out of the reach of most Chinese missile systems. Considering the effective range of U.S. fighters, such a decision would almost invariably prevent American aircraft from getting into the theater. China would have air superiority, while Americans would have to stay away.

The Theater and the Homeland: When Militaries Fight Each Other

If war were to break out between the U.S. and China, their militaries would first fight each other in East and Southeast Asia, most likely far from Taiwanese shores. The larger Western Pacific is therefore the main theater we will need to look at when we try to envision how such the conflict will develop - initially. There, it will mostly involve going after each other's surface vessels, submarines, fighter planes, and missiles, and their logistical support systems.

When people hear about war, most think instinctively of such uniformed militaries pointing and shooting weapons. This is the traditional "kinetic" warfare mentioned above, using explosives or other means of destruction to kill an enemy's military forces and the bases, platforms, and supply routes they use. Notably, during a U.S.-China conflict, the main fighting of this kind would take place either close to the Chinese coast or up to roughly 2,000 miles off the coast of the Asian mainland. This puts it inside and along the Second Island Chain in the area between the Chinese mainland on the one side and Okinawa, Guam, and New Guinea on the other. The theater is so large because Chinese missiles can hit U.S. vessels and bases up to that distance. It is a seriously lopsided situation. Unlike the PLA's Rocket Force, the United States can only seriously engage within a few hundred miles of the Chinese or Taiwanese coast. The range of American planes and missiles is not long enough.

Except for submarines and Kadena Airbase on Okinawa (should PLA missiles not have preemptively attacked it by then), most American air and naval forces could not get close enough to Taiwan to make a significant difference in the actual fighting. To change this, they would have to first prevent Chinese missiles and drones hitting the carrier fleets, and *then* overcome the PLA Navy. Even before surface ships of the U.S. Navy would fight the PLAN, they would therefore need to confront waves of missiles, drones, and fighter planes taking off directly from within China's borders.

Although most Chinese ships probably still are inferior to their U.S. and allied counterparts, the Chinese Navy would have superior numbers and overwhelming support from the mainland. Most likely, only the U.S.' small number of F-22s would have a realistic chance to get close to Taiwan. Other assets could only join in after the main origins of the onslaught have been dealt with.

The United States would find itself in an untenable situation. It could not defend its bases and ships from Chinese missiles. It could not protect its fleet and fighters from being attacked from Chinese bases. It could not interrupt Chinese military supply lines on land. And on the sea alone, it could not significantly enough weaken military supply routes and military manufacturing capabilities.

All these points suggest that America would have to make the probably most consequential decision of a war with China. It would need to attack the origin of the PLA's regional superiority: the Chinese mainland.

Back to the Future: The Battlefield Shifts

Even if successful, or *especially* then, such an attack would most likely result in a massive escalation of the war.

The distinction between American forces hitting missile sites or air and naval bases, and attacking the Chinese homeland more broadly, becomes difficult and maybe even impossible to make. Definitely, the Chinese are unlikely to differentiate and rationalize it that way.

We just must think back to the much more contained attack of September 11, 2001. Or much further, to Japan's 1941 attack on Pearl Harbor, or the Battle of Britain in 1940 and 1941. Like then, those being attacked in their homeland will demand revenge. Or simply retaliation.

The psychological impact would be extreme.

Targeting the enemy's homeland would almost be impossible to contain, even theoretically, and even if the attack would only attempt to destroy military objects - like the Japanese did in 1941.

It would be difficult to define what would be a military target as opposed to a civilian one. What about communication networks, or command-and-control centers? Or supply routes for ammunition and weapons? If it were "fair" to sink PLAN ships delivering supplies to Chinese soldiers fighting in Taiwan, what would be the logic to *not* blow up Chinese trucks or pipelines supplying PLA fighters with the means of attacking American planes and ships in the ocean?

China's military support comes from land next to where the main fighting would take place. A large share of China's military bases is on the coast. This is from where it would launch most of its fighter planes, missiles, artillery, and ships. But others, including bombers, rocket bases, and command and-control-centers, reach far into the hinterland.

Americans may attempt to "only" attack these, flying over a continent-sized country larger than their own. But a billion Chinese could look up and would see the U.S. military do that. What would they be thinking? Let's not bet on them saying, "wow those Americans are serious, let's just give up!" Rather it would be a cry for blood in America.

In the meantime, if the U.S. Air Force would consider hitting land-based supply lines, why not go after some of the PRC's sea-based economic lifelines first? Couldn't this shift the battleground?

Many of the Chinese economic supply lines run through the Strait of Malacca into the Indian Ocean and roughly along its Northern Coast into the Middle East and Africa. This is the "Road" of the Chinese Belt and Road Initiative. But the increasingly strong "Belt" part of the BRI stays on land and far from any coast. It goes west into the Chinese heartland and from there into Central Asia and the Middle East, and north to Russia. And a third, digital, component goes upwards to satellite systems stationed in space.

Most likely, the U.S. and its allies would have no serious issue stopping the oceanic supply routes of the BRI's road, and beyond. They could close the Strait of Malacca, together with all routes through Japan, the Philippines, and most of those further out in the Central Pacific. Geography ensures that China could do little about this. But because of geography's shrinking geopolitical role, it could not protect somewhat comparable actions by China.

Already 80 years ago, during World War II, long-range weapons extended a more traditional concept of a theater of war to include the enemy's homeland. Although the main traditional military encounters took place far away, Germany's bombers and later V1 and V2 missiles brought death and destruction to Britain's civilian population. And after 1942, the allies were even more effective in this kind of warfare once they achieved near unchallenged air supremacy. The German and Japanese homelands became civilian targets for military weapons that flattened and burned down the buildings they lived in and any kind of infrastructure and economically relevant facility.

Bombing an enemy's homeland tries to achieve three fundamental goals that go beyond killing military personnel and destroying their weapons and equipment. It interrupts the other side's military and civilian supply routes, diminishes its source of economic and military power, and distracts and demoralizes its population. The enemy must refocus and redirect critical means to support tasks and to keep people at home surviving and safe.

When fighting close to Chinese shores as is likely in a conflict about Taiwan, many of these actions and goals blur into each other. When we target Chinese military supply lines, the PLA will target our military supply lines as well. These must cross the Pacific Ocean. The PLAN would fight in some ways similar to the submarine warfare during the Second World War, against comparable kinds of targets and weapons. China would not only rely on its larger numbers of attack submarines, though. It would also use flying or swimming drones and swarms of drones,[194] aided by space-based surveillance and AI-powered analyses of images and other sources of information.[195]

But the most dramatic effect would be when the war would involve our populations. This is going to happen at the latest if we were to interrupt Chinese supplies of raw materials, energy, and food, and if we were to bomb positions on the Chinese mainland, particularly the dual-use infrastructure needed by the military and civilians.

And here, China could repay us in kind. It could retaliate and do something very similar to us. To the largest part, it would not even need to use its navy. Instead, and with relatively affordable measures and efforts, it could use our own systems against us. I will explain potential approaches in more detail in the next two chapters.

Our biggest challenge in the fog of war may therefore be to prevent triggering a rapid escalation of warfare. And as we find out that we cannot achieve that, we may also discover that America is not as far and safe from Chinese attacks as most Americans seem to think.

11

Escalation

The World's Most Dangerous Secret: Game Theory Dot Two

"Would the U.S. risk San Francisco and Los Angeles to rescue Taipei?"

Israeli military strategist Martin Levi van Creveld's answer is an incisive counter question: "Do you really believe China will put Beijing and Shanghai at risk in order to seize Taipei?"[196]

Indeed, these seem like the big questions.

But the choices they present are false. War between the world's superpowers can dramatically escalate even without involving weapons of mass destruction. European casualty rates and overall levels of destruction were higher during World War II than World War I, even though between 1939 and 1945 they did not use chemical weapons.

Likewise, even without using nukes, some effects of all-out war between China and the U.S. could be dramatic and horrible for both societies. Unfortunately, most scenarios and almost all public discourse ignore this. Instead, experts in the West often outline three levels of escalation of a war over Taiwan.

On the lowest level are **non-violent actions,** not being formal acts of war. Among these would be American logistical, intelligence, and economic support, sanctions, and maybe even the supply of military equipment and weapons.

Level two involves **kinetic warfare.** This is the active participation of U.S. forces on the battlefield, and probably Japanese ones, too. American planes would conduct dogfights with their Chinese counterparts and battle for air superiority in the war theater. The United States would integrate air power with the capabilities of its submarines and surface ships to take out amphibian landing boats, troop transporters, and other ships. On land, American troops may actively aid the Taiwanese to repel the invaders.

As a result, as the theory goes, either China or the U.S. will consider level three: **nuclear war**. By then, one side will be in dire straits and with the back to the wall.

Some people, like Elbridge Colby, former Deputy Assistant Secretary of Defense and founder of The Marathon Initiative, see the U.S. subduing the Chinese attackers and the PLA paying an ever-increasing price.[197] Either way, one party or both would have to decide on whether to use tactical nuclear weapons. Even if this happens, most consider all-out nuclear war with intercontinental missiles to be out of the picture. Instead, the war would likely grind to some sort of standstill. It would be reminiscent of the standoff after the Korean War of the 1950s.

Another aggression repelled, another war in Asia ending in a tie that both sides can live with. Another "magic" success for strategies devised by what Derek Leebaert calls "emergency men."

Here is the dirty secret, though: All this is a massive simplification. The reality is much more complex and much more dangerous. It may be the world's most dangerous secret.

Consider the following.

From a base in Northern China, a single missile rises into the skies. Déjà vu? Well, not totally. This is neither a "Carrier Killer" nor a "Guam Killer."

This missile is different.

While heavy fighting continues in and around Taiwan and the ROC forces struggle to repel the invasion, the phone rings in the White House in Washington, D.C. It is a call directly from Beijing.

Of course, the president of the United States of America answers. There is usually nothing to lose by talking to one's enemy.

Unfortunately, neither the U.S. president nor the president of the People's Republic of China speak the other's language. They are unable to converse directly, listen to each other, or even just read each other's speeches. All is filtered through, translators, aids, and advisors. This makes it difficult to establish a close personal relationship, or to correctly understand and interpret nuances.

When the American picks up the phone, though, none of this matters. The Chinese president has gone through the effort of practicing three short sentences in clearly comprehensible English:

"You have blocked essential supplies of food and energy and destroyed services needed by our people. We will retaliate. The missile does not carry any warhead."

And then he adds two more words.

"Back Off!"

Before the U.S. president can respond, his Chinese counterpart hangs up.

The missile's rocket propels it far into the atmosphere and then beyond. It slightly passes the Karman line, which science considers the boundary to outer space.

Unlike the DF-21 and DF-27 intermediate-range ballistic missiles described before, this is an intercontinental missile with multiple warheads. It can release several gliders to approach separate targets at hypersonic speeds, anywhere on Earth. In the scenario described here, it heads not to the west but south, toward Antarctica and the South Pole. From there on, only one general direction is possible: north. Within a mere few minutes, it crosses over Brazil and then passes above the Caribbean, approaching the U.S. East Coast.

In the United States, with its forces on DEFCON Level 2, a quick decision must be made. Normally, intercontinental missiles flying toward the U.S. mainland would have to be considered a nuclear attack and may trigger a counterstrike. But this one is slightly different.

The bad news is that it seems to head toward Washington, D.C.

But there only is one missile. The President and the Vice President have been separated, and the Vice President is in the air. Although the incoming missile may carry multiple warheads, it is highly unlikely that both leaders are in immediate danger. Continuity of government is assured.

On top of this, there was this ominous phone call. Of course, it could be a ruse, but why would it be? What could be the purpose of a lie? After all, a single missile can cause only relative limited damage and would not cripple America's ability to respond in the most severe way. Our bombers are in the air, our ICBMs ready to launch, and our submarines prepared to counterattack at a moment's notice.

If it is a nuke, the retaliatory strike at China will be horrendous. Since a nuclear attack is therefore unlikely, the president holds back.

The missile is fast. Very fast. And just a minute later, it begins its descent. Although it keeps adjusting its trajectory, a scary proposition for the U.S. defense, it becomes clear that the target is somewhere in or around New York City. Probably the nerve center of U.S. financial markets in downtown Manhattan? But within a few seconds more, it becomes clear that it must have failed to hit its target. Missing Manhattan, it strikes the ground away from any visible infrastructure or population center, a few miles north of the Bronx.

Did the missile fail? Was it a dud?

Indeed, it did not deploy any nuclear warhead, nor even a conventional one. There was no secondary explosion. But the newest reports show that, indeed, it seems to have split into several sections. These hit the ground relatively close to each other. At the moment of impact, their speed exceeded Mach 5, releasing substantial kinetic energy. As the pieces hit the ground, this must have done something. But what?

And then more information from the target locations comes in from DHS, the Department of Homeland Security. They are five points surrounding the Hillview Reservoir. Two of the projectiles hit locations where the Delaware Aqueduct and Catskills Aqueduct are respectively feeding the reservoir. The others struck three New York City Water Tunnels delivering fresh water to eight million people. One strike by a single hypersonic missile, without needing to carry a warhead, just took out 90% of New York City's water supply.

And the picture is even more dramatic than this. In the meantime, Chinese cyberattacks have delivered zero-day attacks on critical energy infrastructure and the electrical grid.[198] Drone-based cyberattacks and sleeper cells have blown up substations and destroyed pipelines. The American economy is already grinding to a hold.

Because of these attacks on our energy networks, life had already looked dire on the East Coast. It suggested limited electricity and fuel and subsequent food shortages setting in soon after.

And now America's biggest city is running out of fresh water. How can New Yorkers survive? How long can a human live without water?

The above weapon often is called a "kinetic kill vehicle" (KKV). Again, the details matter little. I do not intend to recount what *exactly* would happen, or its likelihood. This is unnecessary because there are too many variations and war is unpredictable. Maybe a missile would hit the nearby Croton Aqueduct as well. Perhaps missile launches would not even be necessary since the gliders, the kinetic weapons, had been put in space even before the attack on Taiwan started.

They could in theory be up there already, in low-earth orbit. This would even give less advance notice. Or it may be more complicated than I described. This is, of course, likely. But in reality, it does not matter whether China would need six independent gliders or require conventional warheads. Or whether it would combine this specific attack with acts of sabotage, or whether drone attacks on some of each aqueduct's 23 shafts would be easier.

It does not even matter whether we can do the same to China.

What is relevant is that it, or something like it, *can* be done. The American homeland is vulnerable to direct attack by Chinese weapons, without requiring nuclear warheads, or even *any* warhead. And the abstract term "homeland" means nothing less than large numbers of civilian American lives.

Escalation Options

For the time being, most countermeasures against such attacks would be highly unlikely to be successful. Like the U.S. can establish a blockade of the Strait of Malacca and the Strait of Hormuz and deny the supply of food and energy to China, China therefore can deny access of food, energy, and even water to Americans. Not to the *continent* of America, but to its *people* living on the continent. We are *that* dependent on complex vulnerable systems. Most of our current defenses against terrorist attacks and their financial networks would be of limited help. A committed state actor with technological sophistication and manufacturing capabilities comparable to our own, a genuine peer, is an opponent several orders of magnitude larger and more capable than a bunch of guys plotting away in the caves of Central Asia or in the tri-state area around Iguaçu Falls in South America.

China can retaliate to a degree far beyond anything encountered in U.S. history. It can do so without requiring nuclear weapons. In the 2020s, the oceans and our navy cannot protect us any longer. Missile defenses are not ready. Critical infrastructure is not resilient. Grids are too centralized. Cyber defenses are too convoluted and therefore largely ineffective. Some cyber tools and strategies achieve the exact opposite of what we need, actually increasing our vulnerabilities and reducing our resilience.

Many scenarios and war games assume that there would be an escalatory threshold that *China* would not cross. But it could well be *us* who would have to shoot first. *We* could be the ones escalating, almost in each instance. China would play defense and deny our forces access to the battlefield. It then would execute a classical game theoretical success strategy: never escalate yourself but always match the opponent's move.[199] Many around the world, including some of our allies, would consider the PRC's use of force against Taiwan a domestic matter. They would disagree with what Beijing does, maybe even despise it. But to them, and maybe merely because it is convenient to look at it that way, it would be the equivalent of a brutal and unnecessary civil war. And therefore, they would not want to get involved.

With this logic, China could decide to *not* shoot at U.S. planes and ships, unless the U.S. does it to Chinese platforms first. China could decide to *not* attack U.S. bases until we launch attacks from them or hit theirs first. China could decide *not* to strike military bases or other targets on the U.S. mainland unless we do so to the Chinese mainland first. China could decide *not* to deny energy, food, communications, and water to the U.S. population, unless we establish a blockade first or start taking out similar infrastructure components inside China.

Each time, China can claim restraint. And each time we make a move, China can retaliate. Step by step. Tit for tat.

The above little fictional story just scratches the surface of it. So, let's dig deeper.

From War Games to Game Theory

American forces lose to China consistently in war-games because of established facts that nearly every military strategist acknowledges. Our supply chains are long where theirs are short, and our weapons and strategies are designed to fight wars anywhere in the world while Chinese' are specifically developed to do one thing above everything else: defeat our military in their backyard.

Trying to prevent the impending defeat of U.S. and allied forces in East Asia keeps us repeatedly coming back to the same option. It is the hitting of air and naval bases, missile launch pads, critical infrastructure, and supply chains of the PLA on the Chinese mainland. We can easily surmise about the likely PRC's response once its mainland is hit. I will elaborate more on that in the next chapter.

The Limits of Game Theory

But such an escalatory step is merely an example of the larger dynamics. Beyond anything specific happening on the battlefield, an additional significant disconnect is likely to appear in our heads. We hear a lot about influencing and directing the other side's rationale. But on both sides, the starting points for psychology, reverse psychology, or a range of game theoretical scenarios are likely going to diverge early on. This further destabilizes the situation.

For China, the ground level of "no escalation" would be its position that even an invasion of Taiwan would be purely a "domestic" affair between the PLA and Taiwan. The PRC says it considers Taiwan an integral, although not integrated, part of their own country. This position means that any action against Taiwan would not involve the U.S.

But if the U.S. considers an invasion of Taiwan already an attack on another nation while the PRC considers it the equivalent of a police action and therefore a domestic affair, both parties would already be out of sync. In this example and interpretation of the interaction, the PRC would consider any U.S. military reaction an *escalation*, an unprovoked attack. The U.S., however, would consider it a *response* to aggression.

Game theory suggests that one of the most successful de-escalation strategy is to make an initial concession, e.g. by refusing to escalate, followed by tit-for-tat responses when the other side escalates. However, if both sides disagree from the outset about who is the attacker and who is escalating, then this cannot work. When either side moves toward a new level of warfare, it would consider it a reaction, while the other side would consider it an escalation.

Rather, and independent of success or failure on the battlefield, such being out-of-sync works like a ratchet or Jacob's ladder. It would function like a mechanism with a built-in feedback loop that favors continuing escalation.

And so, it would be logical to follow a completely rational path toward total war. Similar to what happened during the Great War in 1914. And the man-eater would be at it again.

Homeland Attacks

Simply by looking at this logic, firing at the Chinese mainland would then - of course - invite counterstrikes at the U.S. And China's tit-for-tat response to a naval blockade or the bombing of targets on its mainland would interrupt American supply chains. It could use cyberattacks and break the U.S. fiberoptic, pipeline, and electrical grids via targeted acts of sabotage. In either case, the United States would then have to shift from defending Taiwan to defending the U.S. homeland. Suddenly it would become obvious that the theater had expanded and would now include targets in *both sides'* home countries.

And then America would face the question of whether to leave it at that. Would China and the U.S. see eye to eye on this? The U.S. military and the PLA would each respond according to their own logic in retaliatory moves that seem logically justified by their own audiences.

The specifics are subject to debate. But there are so many moving parts and so many ways of mis-interpreting each other's actions that either side can rationalize almost any decision. These decisions and actions would flow into each other without a clear and natural end point. Both combatants would also speed them up by using AI that, as Kissinger, Schmidt, and Huttenlocher correctly point out, threatens to reward first-strike benefits. All this quickly risks approaching something as close to an all-out war as it can be without nuking each other.

And there still would be no way for the U.S. to win in the short term. A standoff would be the best we could hope for, which would likely be temporary and shift the war into the spheres that we could have fought it in from the beginning: economy, technology, and the new spatial commons.

One can easily see how our complex and highly digitized modern world may render traditional game theory mostly useless. Particularly considering the highly emotional response to be expected when the threat of famines loom in Chinese or North American cities, automatisms could lead to large-scale traditional *and* non-traditional warfare even without the use of nuclear or biological weapons.

And when the civilian hardship increases on both sides, and the casualty count goes up, the use of tactical nuclear weapons may seem like a much more natural evolution of the fight. But when almost all the fighting takes place close to the Chinese coast or even on the Chinese mainland, using nuclear weapons "in the Western Pacific" likely means "in China," or at least "directly affecting China." Unlike optimistic scenarios that assume China would be at the ropes and may revert to a first-use of nukes, why would that be the case if America's bases are destroyed and its carriers and modern fighters could not even get close to China's coasts? It is at least as likely that the U.S. would decide to use nukes to break the standoff.

And then? Do we get all-out nuclear war? Will China, with its lower numbers of nuclear weapons, consider it necessary to start a first strike at America, to prevent nuclear de-capacitation of its smaller arsenal?

Or who will then make the really tough decision and stop the escalation no matter how hard their people were hit and regardless of how much damage already has been inflicted in their homelands? What processes are in place to prevent this from being taken over by the demand of cascading tactical decisions and emotions of decision makers and the public?

In 1914, the localized trigger of war quickly evolved to focus on large-scale operations among the military machines of the largest war-faring nations. Within days, the Great War was no longer about police actions of Austria-Hungary in the Balkans, but about Germany, Russia, France and the U.K. and their respective global empires duking it out in each other's heartlands. Not even World War II started as a great power war in Europe. Nazi-Germany took one step too many against a middle power, coordinating it with its archenemy in Moscow. Within days, Britain and France confronted it, and within just a few years, cities all over the European continent were burning and tens of millions had died.

Likewise, in the 2020s, we could find the war quickly shifting from one about Taiwan to an actual defense of China, the United States, Japan, and South Korea. Likely, the newly formed AUKUS pact would involve Australia and the UK also, turning the conflict into a genuine global one. And any opportunistic Russian action in Europe could pull NATO into the fight as well.[200]

Within a short time, escalation may push all of us into a global fight for the survival of a large portion of the civilian population across our nations. "Short" could mean weeks, days, even hours.

Re-imagining War in a Changed World: The Future Has Arrived

My goal here, again, is not to go into details of warfare. Our governments (we hope) and our potential enemies (hopefully not) have some of their smartest people thinking about and working on these matters. But in the following, I want to encourage thinking outside the box. And this begins by ignoring some harmful truisms and breaking with a few comfortable but outdated paradigms.

Paradigm-Shift One: We Sometimes Are the Underdogs

First, even us having an overall stronger military and long-term advantage will not mean winning a war over regional stakes, and much less any specific battle in such a war. I see little value in proud proclamations of the U.S. being the world's only superpower with the by far strongest military. Such statements are frequently rooted in nebulous historical references to actions not directly applicable to the East Asian theater of the 2020s.

They also simply tend to count numbers of bases or overall numbers of weapons platforms, like aircraft carriers and fighter planes. As these numbers tilt in favor of the other side, vague assertions of a supposed superiority of technology components comfort us. And they disregard way too often how profoundly technology has created new vulnerabilities, for our military and our civilian population also.

We then get a reaction like the one I received in November 2021 from a retired senior-level general. When I asked him about China's likely regional superiority, he responded with the statement, "let's not forget that we always have the option of using tactical nukes." So, going from being denied access to the battlefield directly to using nuclear weapons? I am paraphrasing and simplifying his response, but the implication of the statement was clear. On multiple levels, it was a revealing answer.

It confirms that what really counts is what one can effectively deploy on the battlefield. And there, the picture does not look as positive as it used to be. America's attention and its military is spread across the globe. With a defense budget that may be roughly double the Chinese (but less when considering purchasing power parity),[201] we have significant interests and resources in Europe, in the Middle East, Central Asia, North Africa, and in Asia. Usually, we have one, sometimes two aircraft carriers in the Western Pacific and close to China. Sometimes there are three, but occasionally none.

Moving carriers takes many days if not weeks, even if they are advanced in their maintenance cycles and ready to deploy, or when they are in the vicinity. Apart from strategic bombers, these are our most effective battle-ready means of flexible force projection. Each one can theoretically confront and defeat almost any nation state, except for a bare handful of them.

But China is one of this handful, even if we add the capabilities of our regional bases and allies. China is a peer nation, in many respects much more so than even Russia. This is "peer" in a global and long-term sense. We can assume that we still are more powerful than they are in most regions of the world, and maybe also globally, at least in a protracted conflict. But not in East Asia, and not today. And not tomorrow, unless we draw the correct conclusions about the predicament we are in.

The simple fact of the matter is that according to almost all criteria commonly applied, verified by almost all war games and war-fighting scenarios, and acknowledged by a vast number of our own and our allied military thinkers, we are inferior in the Western Pacific.

Paradigm-Shift One: Understand that we are the underdogs, not the Alpha.

Paradigm-Shift Two: Warfare Will Be Different

Second, the historical way of fighting wars against a peer nation has undergone a profound transition. This one is particularly difficult for many Americans to accept, even to fully wrap their heads around. All our recent wars have used the same kinds of tools, and most of them have followed the same template. We projected power through bases on allied territories and on aircraft carriers, then moved even more air, naval, and land assets close to our enemies' shores. Using our complete air superiority, we then deployed technologically and numerically superior weaponry directly into the battlefield. Recently, this included precision munition, drones, and advanced helicopters and fighter planes.

Throughout the past couple of decades, we have additionally intercepted and surveilled nearly all global data in transit via the Internet or phone communications networks, including financial transactions. We created institutions to limit our enemies' ability to move money and fund their operations. Our data analytics intelligence tools identified their locations so that we could target and hit them with missile or drone strikes, or with special forces attacks. It enabled us to target and blow up our enemies and their equipment almost anywhere.

Boom! That's how *relatively* simple it was for our advanced fighting force.

All this was possible because our enemies were no peers but economically and militarily far inferior nations or ideological tribes, all without modern weaponry, satellites, computing capabilities, or nuclear weapons. The U.S.' global leadership in software, computing, and other technologies like space made our advanced weapons even more deadly and effective than they already were.

Hardly any of this is relevant in the 2020s against a nation like China for a fight taking place in what we must consider the opponent's extended coastal waters.

This would make it not only unlikely that the U.S. could establish air superiority comparable to what, for decades, it had created in any location around the globe outside Russia or the Warsaw Pact. It would also render American supply lines susceptible to interception by submarines, missiles, drones, and space-based weapons.

Some may wonder whether America can balance this with its technological superiority, or through the fact that it has battle hardened forces while China has not fought a major war since World War II. Maybe this is possible, but more likely not. The technological picture is not as clear cut as it looks. Our most advanced fighters and bombers may be superior to their Chinese counterparts. But what good does that do if our fighters' short range does not let them get near the battlefield? And how well are the U.S. Air Force's planes defended against advanced air defense systems that China has built with Russian support and deployed along its coasts? Particularly when our side is grossly outnumbered? And when the supposed greater digital sophistication of our integrated communication and control systems make our most potent weapons also more vulnerable to direct and indirect electronic interference, electromagnetic weapons, blinding of satellite and C3I systems, and general hacking attacks?

A similar logic applies to our still superior aircraft carriers and nuclear-based submarines, and even to our unique ability to perform complex large-scale maneuvers of naval battle groups far from home. Neither of these is sufficiently relevant against well-fortified forces with local support from mainland bases.

We would also need to fight at the enemy's terms, meaning in a location and at a timing opportune to China and not to us. Let's just imagine what us being distracted by an expanding military conflict in the Ukraine could mean for China's plans on Taiwan. And this most formidable enemy ever encountered in battle by the United States would have an at least comparably wide range of abilities to escalate and harm the U.S. economy and civilian population indirectly and through direct acts of cyberattacks and sabotage. In some areas, China's capabilities would exceed those of the United States. I started this chapter with an example of it hitting the American homeland with an intercontinental missile. Alternatively, it could also be one of those conventional hypersonic missiles that is more advanced than the ones the U.S. has in its arsenal. Or long-range or locally launched commercial drones.

Paradigm-Shift Two: Understand that the traditional way of fighting wars is over and that we have neither substantial experience nor an advantage in the new ways.

Paradigm-Shift Three: Vulnerability of the Homeland

Third, and possibly most important, we cannot simply assume that our civilian population is safe. If China considers it an act of retaliation in response to us attacking the Chinese mainland, it could and most likely would attack our homeland and civilian infrastructure as well. And this can have devastating effects that most Americans are not even aware of. Usually, we just hear nebulous references to cyberattacks and an enemy targeting the electrical grid. Sometimes these are mentioned in the same paragraph, insinuating something akin to a ransomware attack on a hospital or a temporary brownout or blackout.

This is orders of magnitude removed from the actual situation. If we want the support of the American people in a world war, we must level with them and explain what really may be in stock for civilians. And this is not something comparable to a California-style brownout or Texas blackout, nor to hurricane Sandy or the Colonial pipeline ransomware attack. Each one of these we can sit out over a few days, with our emergency personnel, army corps of engineers, and DHS working hard. Meanwhile, the rest of the country redirects gas shipments or simply the weather changes.

A common response is that we surely can do to China what they can do to us. This is correct, to some degree. We can certainly hack into Chinese systems and attack their infrastructure as well. But at best this means that we can escalate, or as our own rationale will be retaliate. In its worst case, China still may have more fossil fallback options, less overall dependency on electricity, more hardened electrical infrastructure, and a massive security apparatus to enforce order among the population.

And this is not just limited to the electrical grid. We also must consider our communications infrastructure, data centers, monetary systems, and our current fossil supply lines.

Unlike in the U.S., the Chinese have limited their people's access to the open Internet. They established the so-called "Great Firewall." One side effect is that China knows how to operate when cut off from the Internet. We don't. If even purely a few satellites were taken out and the fiber optic cables connecting us to Europe, Japan, Australia and the Middle East would be cut, it would force our societies and economies, for the first time in decades, to adjust to a world without easy global connectivity.

With all these above actions, China could hit our homeland and hurt our civilian population. As I will lay out in more detail next, successfully targeting strategic nodes or edges of our domestic networks for energy supply, electricity, food, water, and finance could wreak havoc like never before on the American population. Most nations have some mythical imagination of how resilient their populations are when faced with adversity. But not even the Russians, who faced among the most extreme tribulation during World War II, have faced the consequences of a simple act like cutting all communication, energy, and just-in-time supply chains of all goods and products for a modern population of which over 70% lives in large cities.

Paradigm-Shift Three: Understand that our homeland and our civilian population are highly vulnerable to direct and indirect attacks, with potentially devastating consequences for our economy and mass casualties among American civilians.

These three points suggest that "winning" a war against China would be impossible in a traditional sense. They also mean that war is unlikely going to be restricted to some region far away. For the first time in over 200 years, or at least over 150 years, we must be prepared for the actual widespread suffering of civilians within the continental USA.[202]

We have been complaining to each other for decades that America keeps going to war without a clear exit strategy and without a coherent, limited, and measurable definition of success. The most notable exceptions were the First Gulf War and NATO's war in Kosovo.

In a direct fight against China's PLA, though, we are unlikely to have any better chance of "success." We hardly could meaningfully help with keeping Taiwan free. If China fails, then because Taiwan militarily repels an aggression by itself. America's directly engaging Chinese forces would be more likely to escalate quickly and then become extremely expensive in everything we measure and treasure.

Instead of rivals, we would be mortal enemies for generations to come. Many other nations have comparable experiences in their history. In this case, though, it would be a deliberate decision on our side to join them in that.

The effect of the attacks outlined in point 3 above would not just result in the U.S. economy grinding to a halt. It could also mean millions or even tens of millions of American being denied access to electrical power and gas, maybe even food and water, and therefore the most basic essentials in our just-in-time optimized-supply chain society. No food, no heating, no cooling, no cooking, no driving, no medical care, and no access to safe water - for tens of millions of Americans.

The death toll could be staggering. Therefore, let us now look more specifically how this would be possible.

12
Heartland Attacks

Blind, Dazed, and Confused: The Heartland Bleeds (Also)

The new One World Trade Center in New York City, also known as the "Freedom Tower" opened on November 3, 2014. This was over 13 years after the old Twin Towers had been destroyed when terrorists flew two airliners into them.

Rebuilding infrastructure is not just costly. It also takes time. Unfortunately, though, time is not something we would have a lot of if the U.S. homeland were under serious attack by a capable nation state.

If World War 3.0 would start and another great power like China would strike at America, its most potent weapons would not be fighter planes or ships. These would be too conventional and with little chance of success against America's military might. Using nuclear-tipped missiles or bioweapons would be unlikely as well. A homeland attack directly killing masses of civilians would be too crude and unsophisticated to deliver meaningful results short of all-out nuclear war.

The basic approach of hitting the U.S. homeland is completely different, and even comparably simple. The most effective attacks would target America's infrastructure networks. They would trigger widespread cascading failures that deny our population access to energy, food, communications, money, data, and all the necessities of modern life.

Such attacks would be layered, meaning they would use multiple and overlapping vectors that pile on each other. This inflicts maximum devastation and slows down reconstructive efforts. If the destruction were to any significant degree successful, it would take months or even years to repair the damage and rebuild.

In the meantime, and largely without energy, communications, computers, and electronic forms of money, America would stand still. Trucks, cars, trains, and planes could no longer be used. Infrastructure system attacks would wipe out most means to transport people and goods.

Because of our highly efficient just-in-time supply chains, this would then almost immediately also deny large numbers of people access to food, water, raw materials, products and general supplies. It would make it nearly impossible to deliver healthcare and emergency services. Most parts of the economy and of social life, including its most essential components, would grind to a sudden halt.

This is not about stores running out of toilet paper, although that certainly would happen as well. And there is little semblance to today's comparably small-scale cyberattacks that mostly are about extorting a few Bitcoins or stealing our email addresses and credit card information. It is not like asking people to distance socially or work from home, either.

Rather, literally and within days, the well-being and survival of millions of Americans would be at stake. Among them would be most of those about 80% living in the major cities.[203]

Little could be more effective in refocusing American's attention, even from something as severe as major air and sea battles while fighting a foreign power thousands of miles across the Pacific. America and Americans would have to come first, for simple reasons of survival.

Our Modern Society: A Network of Networks

Such infrastructure attacks take advantage of the way our society currently functions. It is therefore worth diving into this a bit deeper.

Over the course of the past decades, and particularly following the emergence of the Internet, networks that connect computing systems have created previously unheard-of efficiencies. Among many more examples, fully digitized systems facilitate and record any kind of commercial transactions. They document property rights and financial assets, pass on orders from consumers and businesses to suppliers, control manufacturing, optimize warehouses, direct truckers and delivery drivers, process payments, and keep records of all related physical and financial transactions.

Most of these systems are connected and create various forms of grids, or webs. This is not a problem in itself, but even mostly positive. Connecting many "nodes" enables us to use software code and AI algorithms to balance over- and under-supply in different parts of these networks, optimizing their overall performance, lowering costs, and improving efficiencies.

So far, so good. However, the current way this works comes with some profound drawbacks. Most of our critical infrastructure systems have in common that they function as interdependent grids with large, centralized components. In some ways, each one of these grids is like the way airlines operate, in a hub-and-spoke model. But then, ice in Chicago or a storm in Atlanta can have severe continent-wide ripple effects across the United Airlines or Delta Airlines networks. Therefore, the just described advantages also carry the seed of vulnerability. By taking out even just a few nodes, the system can rapidly break down.

Enablers of Asymmetrical Warfare

This enables a new form of asymmetrical warfare, delivering far disproportional effects for relatively little effort. It simulates the effect of the just mentioned storms and ice, many times over, to purposefully maximize damage. In its extreme form, attackers can bring down significant portions of the overall network if they take out just a few critical nodes or pieces of equipment inside each of them. They can achieve the same by alternatively cutting out a large enough number of network connections, particularly between the hubs. With the electricity grid, for example, the failure of relatively few nodes (substations) or connections (high-voltage lines) can trigger surges and subsequent cascading failures of the entire system. It then needs to be shut down.

The typical response I get when pointing out all of this is shoulder-shrugging. We all have experienced outages in our lives, during natural disasters, or for a variety of other causes. Hours or days later, or in extreme cases after a few weeks, the power was restored, and everything went back to normal. But these almost always were singular local or regional outages, and the other systems in our country kept functioning. Society could redirect supplies, deliver fossil fuels via trucks, bring spare parts and generators, ship food and water, and clear up the mess.

Normally, we also can cut out the problematic sections of the grid and deal with the problem separated from the rest, while restarting the main part. Of course, this assumes that the attack would not be serious enough to prevent large enough portions of the system to recover.

Either way, though, fixing the specific issues would require most other parts of our country's infrastructure to work. E.g., if an electrical substation is down or a high-voltage line cut, the repair crews must physically be able to reach the attacked locations. This requires fuel. If we only had electric cars, there would be a circular dependency: no electricity, no car, no re-installation of electricity. If we use gas-powered cars, we will require the availability of gasoline. The pipelines, refineries, and trucks then better work.

To coordinate all repair activities, our computers must work, and our communications networks must be available. And, of course, our crews must be able to eat and drink. This means that they need to find and buy food in the stores. But even if all these systems worked, the process still would slow down significantly if enough nodes or key connections were removed.

Our Prime Example: The Electricity Grid

The probably best-known example for such an individual network is the electricity grid. Expressed in one sentence, our electricity system works by generating electricity in power plants, transmitting it using high-voltage lines to substations spread across the country, where it is broken down to be distributed for industrial, commercial, and residential use.[204] In the early days of electrification, thousands of providers competed with independent small grids. Today, all are connected. Most electricity is generated by large utilities at big power plants.

Consumers and businesses use this energy and therefore generate demand, each to satisfy their own needs and without having to understand the large picture. Only recently, they got more directly concerned and involved. Ever more farms, businesses, and households spread across the country generate power. It is for themselves and to feed electricity into the utilities' grid. They typically operate windmills on farmlands or solar collectors on homes and office building.

In the three connected but independently functioning grid systems in the U.S., named the Eastern, Western, and Texas Interconnections, [205] increasingly sophisticated calculations automatically match supply and demand. This led to the designation of a "smart grid." Electricity near-instantaneously flows to where it is needed at any point in time without causing either outages or power surges. This connectedness and a hub-and-spoke structure works with little slack or margin of error. It incurs risks similar to a centralized and mostly hierarchical system. One part breaking down affects everything else.

In other parts of the economy, we have similar networks which form their own critical systems. They optimize their performance by exchanging information about supply and demand. Then they ship the respective materials and products. This is how petroleum, gasoline, and natural gas are provided to those that need to produce electricity or use them to run machines, including automobiles and airplanes. Similar processes also manage the just-in-time supply of raw materials to manufacturers and the supply chains of semi-finished and consumer products, including food.

Domestic Supply Chains

Just-in-time delivery of products reduces warehouse storage space and frees up working capital. As a result, though, at any point in time, a significant part of the products we need within the next few days are in transit. They are not in the supermarket or store, but on ships, in ports, on railroads, or on trucks. Often, they are directly shipped from the manufacturing plant. This makes our society incredibly flexible and efficient, but also highly dependent on the consistent flow of information and energy.

The information directs materials and semi-finished goods as input for manufacturing and to enable maintenance and repairs. And energy produces and moves everything, from food, raw materials, semi-finished products through their often-complex supply chains, to finished products that eventually are sold to us consumers.

The Centralization Problem

In many situations, these networks consist of multiple layers that depend on each other. And unfortunately, today's grids typically are centrally organized and controlled. The size of the networks usually means that if a problem exists in one part, another part can compensate for it. But the consequence of centrally coordinated organization and control is that if we would take down critical parts of the network, or severed their links, the entire system can collapse. And in a hub-and-spoke or hierarchical structure, attacks on the main nodes bring down the rest.

In some ways, it is similar to how everybody else in a computer network can be affected if an administration password falls into the wrong hands. Or how a distributed denial of service (DDOS) attack overwhelms servers with fake demands to a degree that they no longer know what to do and shut down.

All modern activities require energy and electricity, as well as computing and communications. This makes functioning energy and digital networks extremely critical for our society. These systems are even mutually dependent on each other. Without computers and communication, it is difficult or impossible to continue generating energy and electricity. Vice versa, without energy, computers don't work, and we cannot communicate. We then cannot drive either, transport goods and people, and our machines don't work. When energy, computers, and communications are not available, we can do nothing that dependents on them.

If we do not have the energy to run our computers, we cannot conduct most of our business, or process payments. We may not even know who owns what, or where products and materials are at any point in time. Without functioning electrical components, computers don't work either. When our communications networks are down, we do not have information about supply and demand for food, products, or raw materials. Even if our trucks and railroads would work, they may not know what product to ship, and where. And they may not be able to find their way to their destinations.

The Internet - and Money

The *really* complex part about our communications infrastructure is that it consists of multiple layers that all must function together. On the lowest level, it requires hardware, mainly fiberoptic cables, satellites, cell phone towers, and all kinds of little pieces of equipment and their control systems. Above these are Wi-Fi and communication stations, directory systems, databases, and multiple layers of system software code. The end-nodes are our computers, telephones, and other devices. This is usually the only piece we see or pay attention to.

Our financial system requires databases that must communicate and share data to settle transactions. Rarely considered, most government-issued money does not physically exist but is simply a series of database entries to record events. It is centrally issued (a red flag when trying to protect a system), through instructions documented by computers.

A central bank, and above all the U.S. Fed, can create or move practically unlimited amounts of money with a simple entry in a ledger. Again, the availability of such information in databases or ledgers, and it being correct, is critical for the functioning of such a system, and subsequently for our society. Apart from the small portion that exists as cash, all money is virtual and only consists of database entries. We only "know" that it exists because we see numbers in our bank statements, or in money market accounts or pension funds.

For any of this to work, we must be able to trust that these entries exist and are not compromised but correct. To use money, we then need functioning computers and tools for communication. These require electricity, which again closes the circle to the supply of energy.

Food and Water - and Again: Energy

Most of what I just said may sound rather abstract. But when systems break down, at some point things get physical and *directly* affect our survival. We human beings need to eat, drink, find shelter, and move. Only since the rise of the industrial revolution and the relative abundance of fossil energy were we able to solve these problems for almost everybody in our societies. Technology-driven specialization and division of labor improved our productivity. This lets ever more people concentrate on what they do best and offers the tools for them to do so.

The result was urbanization and a system that increased our dependency on food and water supply from distant sources and on efficient storage systems. Like with other products, much of the food we consume is continuously being replenished via deliveries to stores. These arrive every few days, or sometimes, even daily. Perishable food is then stored in fridges and freezers that need electricity to stay cool.

But without communication, trucks will not know where to deliver food and products. Without energy, they cannot even drive. The food stored in fridges and freezers will rot. Our stoves and microwaves will not work. The pumps to deliver water will not work. It sounds like a prepper's dream, or nightmare.

And it goes even deeper. We need energy to grow food as well.

Before the rise of fossil fuels and the subsequent industrial revolution, we had to recycle organic waste to enable the production of sufficient amounts of food. Humans had to manually re-introduce feces containing fertilizing chemical compounds back into the soil so that plants could use them for the processes that make them grow.

In several books and articles, Vaclav Smil, energy and complexity scientist and professor at the University of Manitoba, describes in great detail how this process worked since ancient times and until the Middle Ages. Part of toiling on the fields meant that much of the population, often even its majority, had to transport human and animal feces and spread them across fields.[206]

Today, we artificially create ammonium fertilizers. One of the most consequential inventions of human history, the Haber-Bosch process, enables us to do so.[207] Smil estimates that, without this process, we would need about four times the agricultural surface that we currently use. We would also have to shift our efforts and re-prioritize society's activities similar to what they were in the past. Otherwise, most humans would starve within months or, at the most, a few years. This is also relevant for people living in the United States and in China.

The Haber-Bosch process requires natural gas (a fossil fuel) which it turns via a chemical process into ammonia fertilizer. That process itself also requires energy, which nowadays still overwhelmingly is produced by fossils. It makes this one of the most basic and crude reason for why we still need fossil fuels: we have to eat to gain energy for our own bodies.

All other products and services we use also require energy to design, produce, and ship. And all modern control systems are digitized, in our machines, tools, and factories that make, and in the phones and computers we use to manage them. At the core of all human development is energy. And in a modern 21st century society, almost without exception, all tools and processes require and process information at their core. Our entire society, starting with our most basic necessities of food and water, optimizes productivity and outcomes using data. This triggers an exponentially increasing process of digitization that drives everything in our lives.

All this makes infrastructure attacks so effective.

The Bombing Campaigns of World War 3

For the many reasons just mentioned, tackling our infrastructure can have profound and dramatic results, even existential ones, once war escalates into our homelands. Their cumulative effects on our population and society can reach a scale as consequential as the bombing campaigns of World War II. Very quickly, this can rearrange priorities from shooting at each other's militaries to ensuring the survival of one's citizens.

It is little consolation that the physical worlds continue to exist when the underlying control systems of society depend on the digital realm. When the infrastructure networks are not available, society's functions no longer can be performed.

And this is exactly the goal of such attacks. If any *one* of the networks break down, let alone several of them, the entire house of cards can fall apart the moment the glue dries up. This is, for example, when we run out of stored reserves of gasoline, coal, raw materials, and food. If *all* of them collapse at the same time, it multiplies the effect.

To counter any of this, we would have to ration whatever means and reserves were unaffected. Were we in a situation of war, what would be our priorities? First the armed forces, then emergency and engineering services? Or should it be the sick and the children? Or the young and healthy?

Three of the above-described areas are particularly critical, and they are connected to each other: electricity, fossil energy, and communications.

By taking out even just a few nodes of these networks, each one of these systems can rapidly break down. It is a form of asymmetrical warfare, delivering far disproportional effects for relatively little effort.

So how would attackers go about such attacks, with sufficient means and motivation, and if they wanted to maximize the effects?

Simply put, they would take out each one of these grid networks by going after multiple weaknesses independently. This is what I mean with "layered attacks." It breaks the same network from at least two or three angles at the same time to maximize damage. And then, it also targets several networks at the same time.

Let's look at the various grid systems and how a capable and motivated enemy could do that. These are hypotheticals that do not go into tactical and logistical details.

We will start with electricity.

The Near-Complete Destruction of the Electricity Grid

The electricity grid can be brought down via multiple types of attack, or attack "vectors." Although each individual one of the following may already be sufficient to cause serious damage, for maximum effect a nation state (like China) during a large war would more likely layer them on top of each other:

- Cyberattacks on databases
- Cyberattacks on hardware, mainly chips, firmware, and critical control systems
- Localized EMP attacks
- Commercially available drones delivering payloads on substations and generators, domestically launched
- Sophisticated long-range drones sent from abroad
- Shooting high-powered rifles or rocket-propelled grenades (RPGs) at substations

- Cutting power lines, also likely by using drones
- Attacks using missiles and space-based platforms

First would be massive cyberattacks using software, delivered by insiders and from the outside over the Internet, similar to common ransomware or denial of service attacks. These may leverage back doors built into equipment and software programs that become effective before other attacks, or during an attempt at restoration. Their goal would be to bring down servers and make it difficult to bring systems back up online.

Additional capabilities may already have been introduced stealthily by exploiting "zero-day" vulnerabilities deeply buried in the hardware of critical control systems. Zero-day means that even their existence is unknown to defenders. They could be like those used during the Stuxnet attack on the nuclear facility in Natanz in Iran, sitting on a programmable logic component controlling some of its equipent.[208] Or they could even be buried as deep as on the level of hidden operating instructions physically edged inside of computer chips or networking equipment. Whoever wants to be truly scared, or inspired, could just look at some presentations at "black hat" or "white hat" hacker conferences.

Second, localized electromagnetic pulse attacks delivered via trucks or drones would keep frying electronics equipment and bring down even more of the control equipment and of all tools that use them. The goal here is to physically damage or destroy the inner workings of the equipment. In "Blackout Warfare: Attacking the U.S. Electric Power Grid," Peter Pry describes a wide range of options to use such tools of asymmetric warfare with devastating effect.[209] EMP attacks may target just some strategic substations, maybe in coordination with step three described next. A sophisticated and motivated enough attacker could probably even "short-circuit" the job, although under threat of a massive escalation toward nuclear war. It could use a ballistic missile to explode a couple of nuclear weapons at strategic locations above the United States. According to some studies, this may cover almost all the continental U.S. And when I say "sophisticated enough attacker," I mean North Korea or Iran, not necessarily Russia or China.

Third would be attacks on the grid's hardware, blowing up or otherwise damaging substations, generators, or high-voltage lines. We must keep in mind that these targets are not actively defended. Such attacks can either be performed from abroad, using long-range drones or remotely controlling locally launched versions, or by agents operating on the American homeland. Depending on the formal level of escalation, China may initially rely on acts of sabotage inside our homeland. This may keep the actions below the level of a formal military bombing campaign. Foreign agents that already are present in North America could either fire high-powered rifles, use RPGs, or launch commercial drones similar to those ISIS used in Syria and Iraq.

It does not take many substations to bring down the grid. A 2014 study by the Federal Energy Regulatory Commission (FERC) that then *only 9 substations* would have been enough to have that effect. [210] These were less than 0.02% of the U.S. total of over 55,000. Even eight years later, it seems safe to assume that less than 100 would suffice.

If China would consider its actions as a direct retaliation to the United States bombing targets on the Chinese mainland, they may use military-grade weapons launched from outside the U.S. These could be long-range drones or space-based weaponry. Drones could fly below the radar, fueled by solar panels and autonomously controlled by AI. Heavy enough objects, and even simple tungsten rods dropped from satellites parked in space,[211] could have effects comparable to warheads, without requiring them.

Each one of these activities could suffice. But a combination of several, or all together, would make it extremely complicated to bring back up significant portions of the grid. And the attacks could continue, spaced out and, therefore, also layered in time. As we bring substations back online, for example, enemy action could cut high-voltage power lines in remote locations, with commercial drones.

Why in remote locations? It would be to complicate fixing the damage and require us to move equipment and tools. This may be difficult because so far, this scenario has assumed "only" an attack on the electricity grid. For greater effect, though, similar actions would also destroy pipelines, knock out refineries, and blow up enough of our fuel depots or otherwise make them unavailable. And it would also bring down a significant portion of our fiberoptic network and many of our satellites and disable cellular towers.

Within 48 hours, all shelves in the supermarkets of our towns or cities would be empty, with much of the food rotting because of lacking refrigeration. And few supplies may be forthcoming. Unless you own horses or have access to ranches and chicken farms, you may pretty soon run out of the most essential food items.

Soon, the first people go to bed hungry, in a house without access to air conditioning, a functioning stove, and possibly even water.

And it only has just begun.

Fossils Out: Pipelines & Refineries

Many issues of a failing electrical grid can be compensated, or even resolved, with gas generators and fossil fuels. To maximize the effects of an attack on the electricity grid, a committed enemy would therefore likely target the fossil networks as well. One of the most effective way would be the pipeline system.

Many of us know how that works. Like with power outages, we have seen or experienced simple versions of it in action. A recent example happened in May 2021, with the cyber-attack on the Colonial Pipeline system. Much of the eastern seaboard was without gas. I was traveling down the East Coast through Virginia and North Carolina using an eVehicle. This was lucky. Many gas stations had long lines. In some places, the National Guard was out, whether to keep the peace or to assist with traffic control.

This was just one pipeline system, and the trucks were still rolling. The consequences were manageable, but a single cyber weapon already had a noticeable and widespread impact. Almost five years earlier, a physical explosion in Alabama had a similar effect.[212] About 45% of all fuel consumed on the East Coast touches this one pipeline system at some point.

The May 2021 cyberattack caused serious issues. Now imagine such attacks multiplied many times across the North American continent. The Colonial Pipeline system is 5,500 miles long. It is the longest in the U.S. But the overall U.S. pipeline system is 40 times as big. As of 2020, there were 228,102 miles of oil pipelines on U.S. soil, and 1,647,688 miles of gas pipelines.[213] Can we really protect these all? Keep them safe from cyberattacks by a state with dedicated and extremely sophisticated and capable hacker groups? And from physical attacks through sabotage? Sure, there is some security. But not everywhere and while a serious war is going on.

During Covid-19, we learned that many of us consumers can survive without using our cars a lot. But the trucks must continue rolling and the electricity better works. Our supply chains and our food supply depend on trucking. We also must be able to get to the supermarket or have food delivered to us. Massive interruptions of fossil and electricity grids would quickly have some serious effects on all this. People may literally not get food.

What would happen if hundreds of attacks resulted in dozens of major systems being interrupted?

And it can get worse. Directly tackling our digital and communications infrastructure could be the third crucial leg that an attacking nation state would use to deny our people access to life's essentials.

Cutting Communications & Destroying Electronics

In principle, there are three core ways of data communication: fiberoptic cables, satellites, and cellular towers. Recently, some shorter-range capabilities delivering higher throughput have become the fourth option. Among those are 5G and near-field communication (like bluetooth). But to date, the most important part, the backbone of the Internet and of communication between population and industrial centers in the U.S., consists of fiberoptic cables complemented by communication satellites.

Underwater Fiberoptic Cables

Cutting the transoceanic fiber optic cables reaching the United States would be another non-traditional and even non-violent military response. [214] About 50 of these cables are critical for transcontinental communication,[215] with many of them connecting the U.S. to other continents. Across the globe there are over 130,000 miles of such cables. Many are up to several miles below the surface, and difficult to find.

But they still are crucial for global Internet traffic and considering their typical length almost impossible to protect completely. Much of the debate about cutting Internet cables focuses on the Russians, or on the Taiwan theater. But China has certainly the capabilities to do such harm to the U.S. as well. It would surely consider using them when a military conflict escalates into both sides' homelands.[216]

Were all these cables cut, any transcontinental communication would need to go through satellites. Much of our international communications, including commercial supply chains, would face severe challenges. This would shift the bottleneck to space.

The largest communications satellites occupy geostationary orbits (GEO). This results in some latency, or delays, but also decent throughput. Positioned the right way, just three satellites in GEO can cover the whole earth. They come with two major challenges, though. First, they transmit only a fraction of the overall data being transferred and,[217] second, there are relatively few of them and they are difficult to protect. So-called "killer satellites" can take them out in several ways. More about this in a minute.

Modern, distributed large constellations of thousands of low-earth orbit (LEO) satellites are about to take over a significant portion of global communications. Until then, current satellites will neither be resilient nor powerful enough to make up in any meaningful way for cut overseas lines. Several of such constellations are being established, most notably SpaceX' Starlink. In 2022, these already contribute about two-thirds of all satellites in orbit. As a new system, though, it still is too early for us to rely on them.

But as bad as such measures would be, cutting transoceanic fiberoptic cables and destroying satellites would just be the beginning of what a dedicated and capable enemy would do. Why not go after all communications in the territory of the domestic United States and its allies?

U.S. Fiberoptic Internet Infrastructure

For casual users, which most of us are, the Internet just "is." Hardly ever do we have a reason to think about what happens under the hood and in the background, and what really enables the Internet to work as well as it does. We rarely consider how it is the great universal and equalizing tool that can give new kinds of opportunities to everybody.

The underlying infrastructure, though, matters a lot. And almost all of it, every layer of the Internet, is centralized. On top of it, governments, particularly if aligned with not much more than a handful of companies, can exert significant control over most aspects of it.

But where there is centralized control, there also is a single point of failure and attack for enemies or terrorists.

We may remember who our Internet Service Provider (ISP) is and that ISPs are necessary to access the Internet. It eludes most of us, though, what it means that we in the U.S. often just have one or two options to access the Internet (e.g., using Cable and DISH). And that one company has a near-monopoly on searches and prioritizing the display of search results (Google). Or that three companies control most of the ways we access the Internet (Microsoft, Google, Apple) and two dominate almost all apps (Google, Apple). Two own the overwhelming number of servers and the cloud infrastructure that hosts web sites and databases (Amazon, Microsoft), and a few social media empires control most of our social media communications (Meta/Facebook, Twitter, Google). Notably, the same names of quasi-monopolist corporations keep popping up, and the government has backdoor or even legal front-door access to much of that infrastructure and the data they process.

There are several additional layers beyond the software applications and even the servers mentioned above. This can go as far as the physical enablers of online shopping, the FedEx, UPS, DHL, and USPS. Without these, online shopping cannot work and many consumers don't receive their products. And how many credit card companies are available for online transactions - Visa, MasterCard, Amex, and...?

Consumers, us, mostly are not concerned with this, though. We just deal with apps, browsers, email, social media, and various forms of chat. And then we forget about all the rest.

But an *enemy* during a world war would be concerned, would know all this, and would most likely use these centralized layers to attack the Internet deliberately, for maximum effect. Anything that big companies and governments can surveil and deny to us can, in principle, also be surveilled and denied by outside actors.

The deep underbelly of the Internet is a web of fiberoptic cables. To explain how they work and what it means for a major war with a peer power, I will in the following refer to a study from 2014.

The most critical part of the backbone infrastructure consists of only a few hundred nodes and links. Disabling or physically cutting these would isolate most major population and industrial centers of the United States. The country would descend into hundreds of isolated and disjointed pockets.

For argument's sake, let's exclude satellites, which still only facilitate a small portion of global communications. Cutting the transoceanic fiberoptic cables connecting the U.S. to Australia would mean that Australians and Americans no longer could directly talk to each other or share data. Likewise, doing the same to the cables connecting Boise, Idaho, and Salt Lake City, Utah, would have the same effect on people living in these cities, even though both of them are neighbors.[218]

In each one of these examples, voice and Internet traffic could still be routed indirectly. The data may go from Boise to Reno, Sacramento, San Francisco, Los Angeles, Las Vegas, and from there to Salt Lake City. This then slows traffic down and overloads these alternative routes. Once enough of the cables are cut, even the domestic Internet acts up, noticeably slowing down, and eventually stops functioning.

The cutting of hundreds, or even just dozens of them, could have a dramatic effect. In 2014, a team from the University of Wisconsin, Colgate University and the computer security company NIKSUN conducted a study to analyze the Tier 1 and regional Internet backbone.[219] Among others, they identified 196 nodes, 1153 links, and 347 conduits connecting either 30 miles or more, or connecting over 100,000 people each. The map they created looked similar but even more detailed to the one of the U.S. Interstate highway system. This means it had a length of easily exceeding 50,000 miles.

Hundreds of nodes and links sounds like a lot, but this book is not about a fight with a motley crew of people hiding in Central Asian caves. Instead, we are talking about the means of a major and advanced nation state with over one billion educated and technically well-equipped people. Such a power can take out significant portions of the fiberoptic network inside the U.S. The harm and the impact could be severe.

Satellites

In his 2020 provocative book "Winning Space," Brandon Weichert lays out his scenario for a Pearl Harbor-like coordinated attack on satellites that may blind the U.S. military and deny it some of its most potent high-tech weapons. [220] This definitely must be a concern for the military, considering that several nations (including China and even India) have conducted successful anti-satellite tests.

I don't know all the specifics of how the U.S. protects military satellites. This may or may not weaken Weichert's somewhat sensationalist argument. However, his scenario also applies to the civilian communications network. Now, we are still heavily dependent on the U.S. military's Global Positioning System (GPS) and on communications satellites in GEO.

Today, taking out a few dozen or at most low hundreds of satellites would cripple all communications. GPS uses less than 40 satellites. This leaves us still in a "danger zone" in which it would remain economical for a country like China to attack and render useless these relatively low numbers. Only LEO satellite mega-constellations will change the economics.

In 2020, we passed for the first time the number of 1,000 satellites in orbit. Since then, the overwhelming majority of all newly launched satellites are smaller than earlier versions (therefore called "smallsats") and belong to new mega-constellations. This catapulted the number in orbit to beyond 2,000, with the pace projected to hit 1,700 launches *per year* by 2030. [221] Because of such high numbers, satellite-based communications will become more resilient and difficult to attack, even by a sophisticated state actor.

As with many points I am making in this book, one question people keep asking is whether we couldn't do the same to China. This is true, in most circumstances, and possibly to a comparable extent. The effect is not equal, though, for two reasons. First, China seems to have some advantages with its hardening of systems on the ground, at least as far as communications and electronics equipment is concerned.

And one of the main selling points of the U.S. military is that it would be technologically superior because of a highly sophisticated use of digital battlefield networks. If that were true, then taking it away for both sides would relatively weaken the U.S. position more.

Cellular Infrastructure

The several 100,000s of cellular towers in the United States can be attacked in similar ways as electrical substations. However, it is not quite as simple to generate a comparable impact. By design, cell towers are exposed but not actively protected. But this makes them susceptible to drone attacks. For a profound effect, the numbers would need to be high, though, concentrate on specific regions, or both. This may not be necessary, though, because cell towers need electricity to function. They would already be affected if the grid were down (as would be the devices that connect to them: mobile phones).

Large-Scale Damage and Destruction of Electronics

Above I already mentioned EMP attacks to attack the electricity grid and bring down substations. Of course, the same means could target other electronic equipment, including such at refineries, manufacturing and chemical plants, and any used for communications. Creating just 30 to 50 EMPs via trucks driven close to substation or data centers could have a devastating effect on the national infrastructure. If used in a populated area, their most noticeable effect may be to disable businesses and private household's equipment.

There is some debate about the effects of electromagnetic pulse attacks on industrial and household equipment, computers, and other devices.[222] The biggest questions, however, are not about their principle. It seems to be about how far from the source of the pulse the damage would disable equipment beyond an ability to repair, and therefore require replacing core electronic components.

Other Asymmetric Warfare

Global supply chains involve more than products and raw materials but people as well. At any point in time, many people live or travel outside the political borders of their nations. This adds another factor of warfare to consider.

Hostages, or Protective Custody

During other wars, the incarceration of potential or presumed enemies has been an issue, including during the World Wars in America. But even in the absence of war, China does not seem to be above taking hostages. At least, this is one way to interpret the case when two Canadian citizens were imprisoned in what looked like retaliation for Canada arresting a Huawei executive.[223] After all, China claimed that arresting Huawei's CFO Meng Wanzhou was politically motivated hostage-taking itself. Such measures, even if initially applied subtly, can have an outsized impact.

At least 70,000 American citizens are living as expats in China, and hundreds of thousands of people from U.S.-friendly nations.[224] After the long-term repercussions of the World War II internment of Japanese citizens, the U.S. is much less prone to enact similarly strong limitations for its citizens, and probably even for foreign nationals. It is possible, though, that registration of foreign citizens may be required, and free movements may be limited. For most, this should be easy, considering the near-total surveillance systems we already have in place.

Even such illiberal practices, though, would pale with the likely scale of imprisonment of American and other allied citizens in China. Most of these people are business leaders, academics working at universities, or key contributors working for Chinese companies or affiliates of American or other Western ones. Unlike the U.S., China is not an open society that encourages immigration and integration of expatriates.[225]

To what degree this is an issue depends on how an attack on Taiwan or a conflict between the U.S. and China starts. In case of a sudden attack and subsequent escalation, there would not be sufficient advance warning. It would catch most foreigners by surprise.

Disinformation, Social Pressure & Anti-Trust Campaigns

Another asymmetrical tool of warfare is governmental espionage and classical propaganda warfare. Here is how it could work. The 2015 attack on the Office of Personnel Management (OPM) is a prime example for the first one, espionage. Presumably Chinese government-related hacker groups succeeded with a breach that resulted, among others, in detailed data covering over 20 million background checks being stolen.[226] The practical implication has been grossly under-appreciated.

Every person in the United States seeking a "secret" clearance must furnish foreign contacts, background information, and scrutiny about potential weaknesses in their personal lives. With millions of these records made available to the Chinese government, it delivered a blueprint for how to compromise many people in the U.S. that are in key positions of government.

This data, or similar ones, could also become a pressure instrument to support propaganda efforts. The Soviet Union and its major successor state of Russia have been engaging in large-scale disinformation and anti-trust campaigns for about 50 years. China started some similar measures, at least since the 1990s during its lobbying for joining the World Trade Organization (WTO). Beyond corporate espionage, their goals started out with a focus to positively shape the American image of China.

As Szu-Chien Hsu and Michael Cole describe in great detail in their 2020 book "Insidious Power," this has turned into something much more Machiavellian.[227] China now actively uses its economic cloud to achieve specific foreign policy goals. For this, it deliberately influences businesses, civic organizations, the media, and even universities. China's primary means is to create financial dependencies. One way is to contribute significant amounts of money, which it can subsequently threaten to withdraw. Others are to withhold access to markets after they have become a comfortable and significant source of revenues and profits for foreign organizations, or for non-profits to withhold visas.[228]

Modern Technology and the Heartland at War

The above sounds dystopian. Even so, it is nothing new. Usually, all the points I mention keep popping up in public debates and expert analyses. They just do so almost only, unfortunately, in the context of "terrorism." This insinuates isolated incidences of attack, or at the most relatively contained loosely coordinated actions by small groups of insufficiently funded extremists. And likewise, countermeasures usually focus on that context also.

They do not consider what another great power could do, and would be motivated to do, if war breaks out and escalates. This ignores one of the most devastating consequences of war between the United States and China. It even is one highly probable to happen because of the way actions in East Asia are likely to escalate.

None of the above activities required missiles, fighter planes, bombers, aircraft carriers, or navies shooting at each other. All of them could have been achieved without directly hurting a person. And still, the effects are at least as far-reaching as the most massive bombing campaigns of World War II.

The public debate looks at Taiwan and the Western Pacific, compares capabilities of traditional weapons that mostly are similar to those fighting in the same battle space in the 1940s. And then pundits talk about the likelihood of nuclear war.

The much more probable scenario bringing warfare right into our heartland and to our civilian population rarely enters the picture. The public feels they are safe.

They are not.

And this leaves the big question: what can we do about it?

Part III

GRAND SOLUTIONS

Building and Protecting a Modern Version of the Free World

13

Military Transformation

Realism: A Strategic Pivot to the Future of Geopolitics and Technology

In his "Lost Fleet" space opera, Jack Campbell describes the bombardment of enemy positions on planets using simple kinetic devices.[229] They basically consisted of little more than heavy inanimate objects, like rocks or metal, hurled at planets. Gravity did the rest, with horrendous consequences. In military vernacular, these are nowadays called "kinetic kill vehicles."

A completely different type of weapon appeared in an equally prescient scene of the Gerard Butler action movie "Angel Has Fallen."[230] There, the President of the United States was attacked, and his protective detail skillfully killed by an autonomous drone swarm launched by terrorists. These used computer vision and AI to identify the individual people they specifically targeted.

Both are obviously sensationalist and simplistic scenarios. However, we do not have to suspend too much disbelief. In principle, such kinds of weapons are possible and probably do already exist.[231]

The above examples may even miss important additional capabilities of space, drones, robotics, IoT sensors, and cyberattacks. They also ignore the vulnerability of hierarchically controlled networks, which I described in the previous chapter.

The real world is already today more advanced than the one depicted in these fictitious stories, or at least it is about to become so soon. For that reason, we must consider and confront such weapons more urgently, not just for offense, but also to protect us against them.

"Defense" is an important keyword, because we often focus too much on the more glamorous effects of offense, about affecting others. The bottom line of security is, though, to prevent harm from being done to us. In the mid-21st century, space and digital tools powered by artificial intelligence are critical for this, although in some ways different from how they are currently used.

The key to understand where we are heading with modern weapons is to realize above all that *geography does not matter as much as it used to, while digital infrastructure is becoming ever more relevant.* As seen in the last chapter, the digital can be relevant far beyond cyberattacks or even the military battlefield.

Change, therefore, is not optional. Our weapons and military strategies must adapt.

Satellites, drones, software, and AI are prime examples of what this means. As explained above, China sees them as key tools for its "forward defense" strategy and defeat of American and allied forces. They can become effective and difficult-to-overcome new kinds of weapons, reaching anybody, anywhere on the planet, in a matter of hours, minutes, and sometimes even seconds. It is not only a matter of reach or speed, though. Unmanned objects can be more maneuverable than manned ones. They can also be smaller and operate in swarms, autonomously guided by AI.

All this comes in handy for attackers going after big prey. It does not matter whether they attempt to destroy military objects or civilian targets without active defenses. Their drones, and space-based and digital weapons, can strike at ships, planes, pipelines, electrical substations, or other critical pieces of infrastructure. The targets can be far away in another country's homeland, or ships and planes approaching one's coastline.

On the flip side, the same kinds of tools can also be used defensively. Space-based monitoring and communications, drone swarms with advanced sensors, and AI to analyze all their data feeds can identify and trace physical threats almost anywhere as they appear. Used effectively, no ship, airplane, or missile can navigate secretly.

When such capabilities exist, they will be used. Restraint alone, or a yearning for peace, will not protect us. It would be naïve to think otherwise. Therefore, and although I am asking for non-violent and restrained approaches when confronting China, America must consider the context I described in Part II. And then, it must strengthen its military capabilities, particularly its defense. For this, we must redesign our armed forces' strategies and equip them with the new weapons, tools, and tactics that emerging technologies of the near future enable.

The following four areas outline how. As tempting as it is to describe lots of specific futuristic space-age weaponry, I will leave it at general descriptions.

New Kinds of Weapons

We have plenty of more highly qualified experts thinking about new weapons and developing and testing them. There is a simple bottom-line, though.

The future is about data, AI, robotics, space, autonomous and semi-autonomous systems, and self-coordinating swarms. They will use communications using satellites and ground-based mesh-networks (more about these in the next chapter), and the ever more capable and extensive networks established by the billions of sensors of the IoT, increasingly operated remotely via immersive virtual tools. Because all this is the case, new weapons should leverage these same characteristics. And this is where we must spend our money.

Because it is such a significant shift from today, we need to follow a "first principles" approach. This requires us to disregard our current institutions and tools, at least for a serious thought experiment. The future must dominate our *thinking* already. And this leads us to weapons that are autonomous, unmanned, small, can operate decentralized and in swarms, and tap into rapidly expanding civilian networks of sensors and digital infrastructure.

If we do so, interesting new options appear. For example, we may take a fresh look at the approach of forward defense using a vast worldwide network of military bases, proven over three generations. Today, we are forward-deploying large weapons systems and manned pieces of equipment in specific geographical locations. This turns them and their supply lines into predictable targets which we must defend, maintain, and support. In that respect, and others, our global reach can be as much a negative as a positive.

It establishes "tripwires" that turn our military personnel into targets. They, thus, pull us into conflicts more easily. This also ties up resources and invariably ensures that we will be too thinly spread almost everywhere. Currently, the only way to create realistic fighting capabilities against another great power is then to build up capabilities after tangible threats arrive. It involves shipping weapons, ammunitions, energy, even food, and all kinds of supplies and tools to wherever needed.

This is slow, cumbersome, inflexible, costly, and dangerous. We are unlikely to have weeks and months when war breaks out but. Particularly in the scenario of war between China and Taiwan, it will probably be only days or hours of advance notice.

Virtual Bases, Partly Based in Space

Autonomous, unmanned, distributed swarms of long-range drones may enable us to change this. They could shift many of the current 800 global bases to dynamically adapting virtual bases that can adjust rapidly to changing circumstances. Various types of drones, some potentially based in space, could then be deployed constantly or launched as needed. They could dynamically regroup and strengthen their presence as needed. Such networks can function like kevlar,[232] or honeycomb composite materials - lightweight, self-adjusting, and with powerful emergent properties.[233] They are inherently resilient because they are unmanned and can and should consist of large numbers.

This also requires a shift from human-operated fighting machines to semi-autonomous and autonomous unmanned systems, and swarms of them.[234] Offensive weapons would be semi-autonomous with a human on the loop or in the loop.[235] Other systems can be autonomous, like sensors or smart mines. Satellites and drone swarms can identify threats and follow them without human intervention. For this to work, though, we must shift investments from vulnerable large and centralized systems to smaller mesh networks and swarms.

In Chapter 9, I mentioned that several American military institutions fund and deliver world-class R&D. Too often, however, the issue is what we subsequently implement and put into service. When these are large killing platforms used for heavily debated "forward-defense" wars in countries far away, many civilian researchers balk at contributing to the Pentagon's efforts. This should be different if our primary focus turns to defense and protection. Leading researchers in academia and private industry should then be more open to cooperate on basic research and development in advanced technologies. It can reduce opposition like the one of Google employees objecting to the Pentagon's AI "Project Maven" in 2018.[236]

Military defense requirements mimic the society's general demands and can help further our competitive advantage by cooperating with the commercial sector. This can break down barriers between the military and the civilian sector. On the way, it gets us closer to leverage the full potential of the best that human ingenuity can offer to protect our homeland and society.

Sensors, Drones, and Drone Swarms

When monitoring for physical threats, sensors in IoT devices, and flying and swimming drones or drone swarms can complement the picture generated by traditional satellites. They can monitor specific sea lanes, coastlines, and conduct detailed inquiries when following up on potential threats.[237] Some of today's drones can fly for a very long time (at least several days) and cover enormous distances, operating largely autonomously. Many can also swim or dive.

Uniquely powerful (and expensive) satellites with highly specialized cameras and sensors have their place. However, they are not capable enough. Their "on-the-ground" equivalent should be more maneuverable and less expensive drones, even off-the-shelf ones. Combined with satellite mega-constellations, the can add more flexible and multi-dimensional total data that AI and quantum computing algorithms can analyze. These can identify enemy movements and activities.

Make Military Infrastructure Resilient

The United States also must increase the resilience of its military's infrastructure. The goal is to reduce vulnerabilities and reverse the economics that to date favor attackers. Today, an attacker could limit many capabilities of hundreds of military installations by taking out the civilian electricity grid inside the U.S. This is a serious problem, whether or not done in the way described in the previous chapter. It reduces America's ability to fight a war and help its population during times of emergencies. And as I pointed out above, such emergencies are almost guaranteed when great power war escalates. But if an attacker would have to target each one of these installations separately, the chances for a successful attack plummet significantly.

The below suggestions are equivalent to enhancements of our civilian infrastructure that increase its resilience in the domestic United States. I will dive deeper into some of these in the next chapter.

To continue to function during a great power war, each individual military installation must be adequately protected and have maximum overall resilience. They must be energy independent and have secure fail-over communications systems. Each base must secure the electromagnetic spectrum and, therefore, computers and electrical equipment. This does not just stay on the ground. It also extends to protecting space-based infrastructure.

We can decentralize the national energy and communications infrastructure and distribute its capabilities. Technology has increased the number of tools available for this. For energy, for example, these reach from geothermal energy over microgrids to vehicle-to-grid (V2G) solutions for energy production and distribution. Satellites and device-to-device mesh networks in communications add capabilities and enhance resilience. Multiple versions of fail-over could complement this, like modular nuclear mini-reactors and batteries for energy, and cell phone towers for communications.

Making all military bases energy-independent is a fairly straightforward opportunity with enormous benefits. Objections that this may be driven by an exaggeration of global warming's impact as "existential" are political statements. They ignore today's bases' vulnerabilities. This is about resilience. Military and national security concerns alone demand energy-independent military installations. A positive side-benefit is that they also can protect the environment and reduce CO2 emissions.

For higher energy resilience, we could even explore more extreme solutions in two ways. On the one side of the spectrum, we could create backup and base load capabilities down to individual blocks and even buildings. On the other side are space-based solar power plants that even may shift toward wherever energy supply is interrupted or needed the most at any point in time.

Increasing the resilience of military bases must also strengthen, or "harden," our computer and communications systems from electromagnetic interference and attacks. The Pentagon is aware of this need on the battlefield.[238] With our heartlands exposed to enemy action, it is as relevant to domestic military installations as to the places where the actual fighting may go on. Among others, the Electromagnetic Defense Task Force (EDTF) stresses this.[239] Many of its defensive measures are relatively simple and involve surge protectors and Faraday cages.

Siliconize the Defense Acquisition Process

The slow nature of the U.S. defense acquisition process makes it extremely difficult to effectively take advantage of the rapid change of digital technologies. Military acquisitions keep focusing on incremental, or linear, improvements on 1940s style large and manned platforms. This aggravates structural risks that we can expect enemies to exploit. Decades-long development and deployment cycles, though, can only result in comparably incremental improvements. They easily miss the exponential changes and opportunities driven by innovative quantum leaps and technological breakthroughs occurring within just a few years. Most of these leverage the continuous doubling of performance in computing and communications.

To overcome this deficiency, we must shift toward shorter and more frequent development *and deployment* cycles and introduce more experimentation with innovative concepts and capabilities. This will help us shift more radically toward where the world is heading and not where it is right now, or where it was a few years ago.

Many business books keep describing how inflexibility and sticking to their huge past successes abruptly ended industry leaders' market dominance. Prime examples are photographic film maker Kodak and movie rental franchise Blockbuster. Ignoring the impact of new technologies, digital photography and movie streaming, they stuck to outdated tools and business models. Once adoption of these new technologies reached a tipping point, both previously highly profitable companies lost their market-dominant positions within a matter of a few short years. It is a warning that should ring in every politician's and military leader's ears.

This is only part of the story, though. Often forgotten is that large and important actors in each organization had understood what happened. Not only were they prepared to take aggressive action, but they even recommended it. Kodak's research teams were even leading the world in digital photography technology. Blockbuster's CEO, after initially having rejected the opportunity to cooperate or even merge with startup Netflix in 2000, soon after reversed his opinion. Once he had properly processed the impact of the trajectory of new technologies, their devastating effect for his proud but outdated business model had become obvious.

The decision makers did not get it, though. Kodak's R&D department was ignored. [240] Blockbuster's CEO John Antioco eventually was fired by the board and replaced with Jim Keyes, who oversaw the company's eventual demise.[241]

Today, we have a similar challenge to digest the full impact of technology on the military. Probably nothing I describe in this book, particularly anything related to technology and weapons, is unknown inside its branches. The U.S. armed forces' R&D arms are world-class. Many of its thinkers and leaders understand technology and geopolitics extremely well and keep proposing new weapons, strategies, and approaches.

But the money still prioritizes the old and familiar. It keeps focusing on military offensive capabilities far away from our homeland, and on using large platforms and manned killing machines.

The equivalent of the supervisory boards, the defense appropriation committees and politicians outside the military (and inside!), keep sticking to what they know and like. And these are better versions of the old. They look at moments in time and realize that just a couple of election cycles ago the world seemed simple and clear. But this same world is changing exponentially, making *trajectories* more important than the absolute improvements.

If one would add each year 10% more to the effectiveness of a weapons system while the alternative approach doubled its capability annually, the effect soon would be huge. Within 10 years the "new" would have become 1,000 times as powerful as before, while the "traditional" would be less than three times as good. Equipment that was, however measured, at the beginning of a decade only 5% as effective as the old tool would be about 15 times as effective at its end. And, therefore, 10 to 15-year procurement cycles no longer work.

It is deceptively easy to miss this mathematical logic and what the exponential growth of digital technologies really means. Or maybe decision makers understand, but don't know how to explain it to their constituents. Either way, *this must change.*

As a critical means to achieve all the above, we must shift the development and acquisition process for weapons and related tools. It must allow for massive experimentation and rapid deployment of new capabilities. The terms "experimentation" and "deployment" are connected. In Silicon Valley, as in other technology hubs, this has become the standard. It is also called "being agile" or to "fail frequently and rapidly." Its underlying idea is that even failure is feedback, and feedback is good. The faster and more often we get it, the more we know, the more we learn, and the better the ultimate outcome becomes. The most successful businesses of today use mostly self-organizing teams without traditional hierarchies. They constantly interact with customers and gather their feedback after each multi-*week* "sprint" has delivered some new capabilities for a constant work-in-progress.

Such a process delivers progress much more rapidly. And it is possible for the military as well. Just like in the civilian world, the military also must place solutions into service much more rapidly. Its personnel must learn, build institutional knowledge, and habits that quickly adapt to continuous change. All this can be done. Most of the new kinds of systems and weapons just mentioned will probably be small and nimble, and therefore also relatively inexpensive. Again, this is a way of introducing resilience into the process. It opens doors for ever more businesses to cooperate with the military. This means more competition, more nimbleness, more tapping into the diverse creativity of the minds of our best and brightest.

The prospect of receiving $10 million or $20 million for a year or two of development can be a huge motivator for a startup. This is where "dual-use" gets a new quality. *Defensive*, more *efficient*, and more *resilient* capabilities improve all of society. Newly developed systems often have military and civilian applications, and many of their components come from the civilian world anyway. This turns the military both into a driver of R&D for all of society (while keeping its technological edge) and a testing bed for wider applications. The border between "defense industry" and "technology industry" becomes blurry. Test and feedback loops constantly improve solutions, production numbers increase, prices fall, and the before-mentioned Wright's Law speeds up benefits.

The top-down long-cycle mass-producing days are becoming extinct. They always had their limits. Already Napoleon Hill's bestseller from 1937, "Think and Grow Rich" pointed out two corresponding facts. [242] First, most inventions required their creators to envision something that the public did not see because they did not know it. Instead of cars, people would have wanted "faster horses." The 21st century military equivalent would probably be "stealthier planes" and "bigger ships." Second, Hill stresses the need to not overthink things but to get products out to the market fast and test them out by gathering responses from customers.

With today's technologies, these suggestions are not overly difficult to achieve if we simply put our heads to it, will learn from the civilian world, and open our minds. We just must get more comfortable with failure and welcome it.

Focus on the New Digital Commons

A war against China can hardly be won, particularly not in the sense of a World War II-style total capitulation of the enemy. For that reason already, our thinking must focus on what we definitely *can* achieve, which is defense. And there, besides the American homeland and those of our treaty allies, we must above all defend the new commons of the digital and of space.

In Chapter 7, I outlined where the world of technology is heading. The new commons of the digital world and space will increasingly dominate this future. We will no longer depend on shipping semi-finished or finished products around the globe as during the 2000s and 2010s heydays of physical supply chains. And therefore, we must actively commit to put our focus exactly there and not where the world was in the past.

As a logical consequence, freedom of navigation operations (FONOPs) are becoming an outdated concept. Focusing on the "commons of the sea" and therefore the ever less important physical shipments of products is looking at the world through the rearview mirror. If it is not already the case today, the near future definitely will see the most sizable portion of trade being conducted digitally. We can measure this by looking at the value of purely digital products (e.g., entertainment, design, or educational services), plus the digital contents of physical products (e.g. in automobiles, robots, computers) and their components.[243]

In East and Southeast Asia, there is another peculiarity that I had hinted at before. Most of the trade in that world region is trade from or to China. Wouldn't it then be logical to characterize FONOPs as mostly trying to protect Chinese trade from Chinese interference? FONOPs consist of sending naval vessels near islands and territories claimed by China. If the internationally accepted legal status of just a few islands would change, quite a few of these areas would become off-limits.

This could happen with a mere decision and agreement among nations. The logical consequence is that too much about FONOPs seems great power grandstanding rather than tangible defense. It risks the lives of our sailors and airmen and makes our allies and friends nervous. We show up provocatively, without realistic ability to successfully follow through if events get out of control and escalate, while risking exactly such escalation, up to war in an entire region. U.S. presidents and administrations talk a lot about shifting the American focus to Asia. Then, though, they show neither a sufficient economic nor military commitment to anything south of Okinawa. What is it really that we want?

We must look ahead. And the future clearly is about global integration of virtual supply chains, electronically transmitted digital designs and digital twins, and products physically fabricated near-shore or even locally.

These are the new commons.

Shift to Protect the Digital Worlds of Web3

Therefore, the U.S. military must vow to protect space and digital commerce, not top-down but by strengthening the ability to safely and freely own and exchange data and digital assets, share knowledge, and conduct commerce. We also can make this system available to all people on the planet. The means to do so are technical platforms. For how to do it, we can orient ourselves on the success and failure of the Internet.

Although kicked off by governmental institutions, the Internet's most impactful civilian and commercial use was based on a decentralized and liberal philosophy, not on a top-down government-centric approach. There, the United States played an indirect but crucially enabling role. Its legal system protected Internet governance. For a long time, this made possible a more open and decentralized infrastructure than the Internet most likely would have been otherwise.

Over time, and particularly since the 2010s, though, governments and what former Wall Street Reporter and futurist Amy Webb calls "The Big Nine" (Amazon, Google, Facebook, Tencent, Baidu, Alibaba, Microsoft, IBM and Apple) started bending the Internet to serve their goals of maximizing advertising revenues and sell products.[244] This then morphed into the ability to predict their customer's human desires, actions, and thoughts, and to shape them. The "Big Nine" all began using aggressive tracking and surveillance as means to this purpose, amassing unprecedented power in just a few organizations, which all are more or less closely cooperating with their respective governments.[245] This directly challenges a free and liberal world.

We can and must ensure that the emerging new infrastructure of the Web3 will be a liberal and decentralized one closer to how the Internet of the 1990s operated than the one of the 2010s. This will be a clear and profound philosophical counterpoint to how China and other autocratic systems like to operate. Top-down controls, though, is even what many people and governments in the so-called liberal or free world seem to really want. They say they want freedom, but demand and use solutions involving heavy-handed top-down control, surveillance, censorship, and ever more regulation. Now that the Web3 is being established as the new Internet, we must spring into action. We do not have to shape specifics, but we must ensure that Web3's underlying infrastructure and protocols are compatible with a rules-based liberal global system that is free, decentralized, and empowering people.

This is neither theoretical nor just for "techies." Such a system must be *designed* to be secure and protect individuals and their actions, data, and other assets. In some ways, this is like the role of the navies of the past. Exemplified by the NSA and other intelligence agencies, governments must therefore shift their focus from hacking into systems and using universal surveillance.

Together with the R&D and investment arms of the military, they can work together with academics and private businesses to ensure that we architect the modern infrastructure of Web3 to be secure and un-hackable. I will explain more about it when describing a future-oriented liberal world system in Chapter 15 of this book.

The NSA and other military intelligence and research institutions have the skills and funds to play a major role in this. With the right mindset, one that is defensively oriented and fully committed to liberal principles, they can reach the hearts and minds of our own people and experts in the civilian communities. It could turn into a more attractive private-public partnership for many that today are reluctant. The military is more appealing if it focuses on protecting, defending, and strengthening our societies first, and not on killing, surveilling, or controlling others and our own people.

Again, though, I am not being naïve here. Physically and kinetically fighting is obviously one of the core tasks of a military. But the most successful military is one that never has to put its abilities to kill and destroy to the test.

Encryption and Blockchains: Protect our Infrastructure

To contribute to protecting this emerging digital infrastructure for the world, the U.S. military must shift part of its priorities and activities. Traditional intelligence and a focus on physical weapons platforms must move toward strengthening the new digital platforms. This is defense of valuable infrastructure and people, completely in line with the mission of the U.S. Department of Defense: *"to deter war and ensure our nation's security."*[246] It means building, or "baking," security and protection into our systems and data. This requires a firm commitment to work with the civilian sector to secure all systems in our society, even from the prying eyes of Big Data and our own governments, *including the military itself*. The way to do it will likely require encryption and the use of blockchains or other DLT technologies.

Both these technologies should become core instruments of defense. Intuitively, many readers may wonder what this has to do with tanks, missiles, guns, or planes. Indeed, the answer is "hardly anything." But in 2021, the U.S. spent over $84 billion on intelligence, $61 billion for "national intelligence" and $23 billion for "military intelligence."[247] On top of this are many billions for basic research, tens of billions on applied research and development, and over $70 billion for the Department of Homeland Security. This puts the overall dimension close to China's total 2022 defense budget of about $230 billion.[248] If a significant portion of this number would be spent on encryption and governance that makes society safer from the prying eyes of any government, it would contribute majorly to prevent the gloomy consequences of the scenarios laid out in the previous chapter. Today, the dominant thinking still is "we protect you," instead of "we help you protect yourself." Even more powerful would be "we do not need to protect ourselves because the systems' design makes it unnecessary."

Defense and security must come first, not surveillance and law enforcement. Today's cyber defenses cause a philosophical dilemma and lead toward a logical dead-end. They assume that some central agency, or artificial agent, must access all data and surveil all communications of all people and institutions as the only proper way to secure a free world. It means giving up our freedoms in order to be free. This is the same logic used by authoritarian rulers and societies since the dawn of humanity to justify never-ending conquests and demand absolute powers. Such logic can only lead to an untenable outcome and is inherently incompatible with a genuine liberal worldview.

It is *China's* argument for controlling its society, creating security, safety, and establishing order. And even before 9/11, and particularly afterwards, it has been and still is the dominant way of thinking in America and most other parts of the West as well. But the above OPM hack is an example of what happens when centralized systems then give the attackers an asymmetric advantage.

Our continuing inflation of sensors and devices increases the complexity of cybersecurity tasks exponentially, in a way that traditional approaches can hardly cope with. All the while, attackers only must get through once, with ever more possible angles of attack. The more devices come online, the more they get integrated, and the more they get automated, the larger the number of potential weaknesses. It gets, relatively speaking, easier for the attackers but more costly for the defenders. And we are just getting started. Soon hundreds of billions and up to one trillion of IoT devices will come online and even be woven into our clothing and consumer products. AIs will write software code in ways that humans will have no realistic chance of understanding.

This makes traditional approaches uneconomical, maybe even impossible. Many of us keep spending ever more money and effort at home and at work on ever more complicated cybersecurity tasks that add zero value to what we really want to do (i.e., business, education, socializing, entertainment). Despite that, we are subject to even more failures, hacks, and losses than ever before.

We can re-imagine digital defense. Cyber defense always has been a cat-and-mouse game whose economics favored the offense. We can turn this around.

The New Role of Space

The U.S. military must also review its approach to space. This process made a major leap with the recent establishment of the U.S. Space Force. Space technologies can help to harden and protect communications and monitor infrastructure and adversaries' actions.

One of the most potent measures in this respect would be to pull a portion of radio frequencies from national influence. More about this also in the next chapter.

Currently, many of our space capabilities are about communications and monitoring physical activities on the Earth's surface and oceans. We must continue this ability with a mixture of space-based military radar and optical satellites to track weapons systems. But we also can extend capabilities of traditional satellites by combining them with (a) large numbers of civilian-style cubesats or other types of micro-satellites, (b) ground, water, and air-based sensors and drones, and (c) AI and machine learning tools to analyze the data feeds.

Of particular interest is the ability to trace not just movements of surface ships but also of submarines in great depths below the ocean's surface. The analytical capabilities of AI can detect near-imperceptible noise and wave patterns and propose further actions. Identifying submarines capable of delivering a nuclear second strike could have a dramatic impact on competition between the great powers. It could be similarly disruptive as an effective ballistic missile defense.

The general goal with space and drone-based monitoring is, though, to identify actual and potential threats from traditional weapons systems. These are the kinds of weapons that most current scenarios assume would be used in and around Taiwan or along the coasts of China and our Asian allies. They are ships, planes, and missiles.

Such monitoring could center on defense corridors and specific locations, like the sea lanes between islands and archipelagos, and to detect physical threats near our borders. It would eliminate most of the geopolitical benefits China seeks by establishing control over Taiwan and the First Island Chain. When China cannot get such benefits, it then lowers its motivation for military action.

Changing the Nature of Work at the Military

In 1920, 25.9% of the American labor force worked in agriculture.[249] Just 100 years later, in 2020, it was below 1.4%.[250] In 1970, 26.4% of the American labor force worked in manufacturing.[251] 50 years later, also in 2020, the number was 8.5%.[252]

In the meantime, the output of both agriculture and of the manufacturing industry multiplied, and the country became vastly wealthier. Throughout that time, our lifespan expanded from 45 years to 79 years, our work hours shrunk, and even most in the bottom quintile of today's population have access to tools of leisure and comfort that would make the nobility of 200 years ago blush with envy.

This came with a price, though, or rather a challenge, to continuously learn, change, and retool. It took a deliberate effort for most of the workforce. For their own and their children's benefits, new generations moved from often literally being "rednecks" because they physically labored on the fields in the heat of the sun, to being "blue-collar" because they physically worked in mines and factories, to being "white-collar" because nowadays 70% of the population works in a professional or office environment. This challenge took place over decades. It worked, though, and hardly anybody would really want to go back to the old times.

It is not a question of whether we *like* these underlying shifts. But to increase prosperity and advance, we cannot stand still. We only can change where we are if we are moving. Therefore, those that proactively take advantage of the new opportunities will advance. They get wealthier and gain power. The others will be left behind, as history shows about the countries, businesses, and people that did not adjust and change when the world did.

This applies to the military as well. And therefore, our military and their leaders in civilian life, including politicians, thinkers, and academics, better get on board. In 21st century terminology, the military must do the equivalent of *moving from being blue and white collar to no collar*. The alternative is not pretty, as Part II of this book hopefully described sufficiently for the context of a war between the great powers.

Adjustments must become part of the military's DNA because the pace of technology-induced change is accelerating. We no longer have the relative luxury to adjust over decades and generations. Today, it is all of us who continuously must educate ourselves and constantly must adapt.

Already in the very near future, the greatest hero no longer will be the top gun in the fighter's pilot seat, not even the one sitting in a $100 million machine with awe-inspiring situational awareness as part of a well-oiled and smartly managed global machinery. Rather, it will be the AR goggles-wearing geek sitting thousands of miles away employing AIs and quantum algorithms to make best sense of the feed from zillions of smart sensors operated by swarms of autonomous systems in the water, on land, in the air, and in space, as part of a decentralized and self-organizing world.

Sorry, Tom Cruise.

14

Homeland Resilience

Defense: Future-Proving Our Society

Miniature drones that look like birds are observing everything wherever people congregate. Cameras are watching most public spaces. They are recording and reporting any kind of activity in real-time, storing all in central databases, for AIs to analyze and respond. Using facial recognition, all people are followed and their interactions with all others are documented. At home and at work, tools analyze every word people utter, document every song, news item, podcast, and movie they consume. They analyze the contents of emails and chats, and track and store all online activities down to the scrolling, mouse-clicks, and websites visited.

All physical and digital actions of all people are constantly being surveilled. They are then evaluated and judged based on the level of each person's conformity with a bureaucratic and technocratic class's vision. Based on this analysis, people's freedoms and opportunities are granted or limited. It affects their ability to rent cars or buy houses, the price they pay for such and online purchases, the education they can get, and the jobs they can perform.

To this purpose, public and private organizations, governments and some of the largest businesses in the world work hand in hand to aggregate as much data as possible about each person and empower artificial intelligence algorithms. Their goal is to create better products, keep people safe, uphold laws, and to make society more efficient and productive. More data means AIs can anticipate wants and desires, optimize solutions, identify dangers, improve services and products, and create a more just society.

In "The Big Nine," Amy Webb paints the above picture to describe the extent to which China leverages AI to create a totalitarian technocratic society. This is scary for the casual observers and readers who value their freedoms, and it should be. For most in the West, it is a reason to oppose China's totalitarianism.

But all this is possible, and most of this happens to a large degree in the U.S. and other Western nations as well, as Webb correctly points out also.

Almost all words so far written in this chapter could have been about the United States or other Western nations. In America and the West, most of these actions are not (directly) done by the government. Many are neither as consistent and centralized as in China.[253] Also, a liberal legal system frames and partially checks them. But technically, they are comparable, even in their extent. And in some areas, they have similar consequences. We also have extensive physical and online surveillance and tracking, the swiping of our data by quasi-monopolist super-enterprises, the exploitation of this data via digital fingerprinting, cross-referencing, data analyses and use of AI algorithms. This also leads to judging, censoring of opinions, restraining of speech, and freedom-limiting AI-generated scores.

Freedom can already be limited by such simple but highly effective means as sorting search results or "personalizing" what products or sources of information and news are being presented. Increasingly, public-private partnerships and initiatives in the West give government agencies access to all data, subjugate businesses to political priorities, and prioritize thoughts and ideas developed by small groups and "elites." All this establishes both antagonistic tribalism and conformity-inducing social pressure.

Harvard professor Shoshana Zuboff calls the American version "surveillance capitalism" and made it the title and topic of a 2019 book.[254] What this means is governments and big businesses using 21st century technologies as tools that would look familiar to those having lived in the totalitarian societies of the early and mid-20th century. Then, just two and three generations ago, Western democracies had firmly stood up against such practices. They sacrificed millions of lives and vast economic resources for this goal, on European hills and Asian islands.

Today, the record is much more mixed. Too many among us, particularly among those highly educated working in bureaucracies, politics, academics, and business, seem overly confident that they know what is best and right. This lures them into ditching the messy bottom-up trusting in ordinary people's pursuits and judgements. Instead, they excuse their use of technology in a top-down fashion.

Even in the third decade of the 21st century, many people find it difficult to dissociate themselves from such old-style, centralized, one-solution-fits-all, hierarchical thinking. And so, they fail to see the profound disconnect between the growing use of powerful tools with totalitarian capabilities on the one side, and our theoretically free, diverse, and decentralized political and economic system on the other.

All this is highly relevant if we want to protect our homeland because our very first step would require taking away this ability to centrally surveil, manipulate, and control. It would shift ownership and governance over data down to individuals and small communities. The technical goal would be to introduce self-organizing democratic and market principles into the parts of our society that are most important for our future.

That is no slight change from how the world works today, in most societies. At the beginning of the 2020s, we are still using 20th century-style control mechanisms to check digital tools whose powers, dangers, and limitations we only have just started to understand. I will elaborate more on this in the next chapter.

If our thinking remains hierarchical, centralized, and top-down, though, our society remains particularly vulnerable. My number one premise of this chapter is that the U.S. society needs more resilience. And the best way to achieve this is through the distributed, decentralized, self-organizing, and bottom-up organization of America's most critical infrastructure. This applies to computing and data, communications, energy, and supply chains.

Resilience: Diversity and Decentralization

Protecting our homeland against the kind of infrastructure attacks described in Chapter 12 requires resilience. This means above all two related principles: *diversity* and *decentralization*. They directly counter the above-described layered multi-dimensional attacks that enemies can conduct today. Diversity and decentralization change the underlying economics and shift the asymmetric advantage from the attackers to the defenders.

Diversity means pursuing multiple approaches in parallel and creating several layers of redundancy. But the redundancy of sources of energy, like geothermal, solar, fossil, and nuclear, is not just an example of diversity. It also is a form of decentralization.

Another one is the separation of networks into smaller geographical or political units. We can and should do that for each political entity like a city or county, or a larger geographical area. No longer can attackers then assume that they merely have to crack *one* code, or bring down *one* system, from anywhere, to trigger the breaking down of the overall infrastructure across a whole continent.

And it can even get more detailed. The internal distribution of control over each network can be decentralized. Distributed control systems add another dimension of resilience because they eliminate single points of attack.

An example shows what this would look like. We can use multiple sources of energy for electricity, e.g., solar, geothermal, fossil, and nuclear. Adding multiple ways of distributing this energy, e.g., using vehicle batteries, hydrogen, microwaves, power lines, would make the distribution system diverse and create redundancy. Dividing it then into smaller networks, microgrids, for example organized by states, regions, cities, or neighborhoods, would strengthen resilience even further.

At that moment, there would neither exist a single nor a simple attack vector anymore. This would make it complicated and uneconomical trying to bring down all of them for any community. It would be more challenging if we would decentralize the control systems *within each of these electricity networks*. There would no longer be single points of attack. And this works. Blockchains are one example. The financial industry uses them to agree on and then secure their data. The overall effect is that cyberattacks, and even physical attacks, become orders of magnitude more difficult and costly.

An attacker would have to overcome most of all nodes in each distributed system and across all diverse sources of energy. In most scenarios, this level of resilience would make attacks uneconomical and very difficult to conduct, even in each individual city. The challenge would be exponentially larger if the goal were to bring down the infrastructure in all the (then) thousands of independently operating and self-sufficient locations of an entire country. It would require a massive, complex, and nearly impossible to coordinate operation. The costs would skyrocket despite vastly lowered chances of widespread success.

Computing and Data

Digitized societies like the U.S. are obviously highly dependent on computers and devices. After all, nearly all bank and property records, documents of business transactions, health data, passwords, and so much more exist mostly or only in digital form. This is continuing to expand and engulf nearly all aspects of the economy and society. Therefore, we must harden our electronic equipment against damage, including from EMPs.[255] I mentioned such physical protection of electronics in previous chapters and will therefore not elaborate further.

But keeping servers and databases secured from physical harm is only a means to an end. The real objective is protecting society's data themselves. We must keep them safe from prying eyes. The data's integrity must also be guaranteed. This means making sure that they have not been tampered with or deleted, or that such tampering cannot happen in the first place.

It leaves us with having to protect our data and activities from hacking, destruction, and manipulation. The overall goal is to secure privacy and access to our homes and workplaces, and the devices we use. These may be industrial control machines, robots, cars, medical implants, computers, phones, or any other kind of IoT devices. And, above all, they are the data we and our devices create and use.

Building resilience for our data works in some ways like the above example with energy. It requires decentralization and eliminating single points of failure or attack. To protect the homeland, we must encrypt and store all important data in a way that no such points exist.

Encryption of Data

Demanding strong "censorship-resistant" encryption has a poor reputation among some in law enforcement. This is unfortunate because it does not intend to discredit services that protect us, nor to prevent them from doing their job. Rather, it is about better outcomes: more security. Doing so can make law enforcement's job *easier*. Surveillance is only possible if systems are vulnerable and can be accessed from the outside. Protecting them from everybody reduces unlawful activity.

For example, on July 6, 2022, the U.S. FBI and the British CI5 issued a joint statement warning of China's governmentally sponsored hacking activities. Our instinctive reaction should want to protect our systems to prevent such attacks. Too often, though, it is to grant the FBI and CI5 more powers to chase down hackers *after* they stole intellectual property and corporate secrets. But it should be the other way around!

Data and secrets should be secured, and centralized powers reduced. And such defense-focused priorities are even more important if another superpower protects these perpetrators. Often hackers act on another nation's behalf and on their territory, which puts them effectively out of reach of Western law enforcement anyway.

America should not base its defense on insisting that its citizens' and businesses' data should stay vulnerable in case something bad happens. How can one keep a straight face when asking for more surveillance powers to prevent hacking? It demands easy access to data to supposedly make access to that same data more difficult. Say what?!

When the U.S. OPM was hacked, described above in Chapter 12, most likely by China, the hackers got full access to highly sensitive and detailed records of over 20 million people. Getting to this data was difficult. It required sophisticated skills. But the attackers also only needed *one* single hack. Would each record have been encrypted and access controlled via a distributed mechanism, it could at least in theory have required over 20 million hacks. Centralization made us particularly vulnerable. And this is even the case when we assume that 100% of all those that are on "our" side always are and will stay "good," never misusing their powers. One does not need to be sarcastic to conclude that this is not the case!

All this becomes even more important as 5G and 6G applications, peer-to-peer communications, the Internet of Things, and autonomous systems take over ever more aspects of our societies. In a world dominated by digital technologies and data, our first and most crucial tool of digital defense is encrypting this data and any data transfer, and securely governing access.

This is a high-impact opportunity for the NSA and military to co-develop quantum-secure encryption together with academics and industry, and then promote its use by businesses.[256] There are many tens of billions of IoT devices, 5G communication nodes, and edge computing tools to protect, and soon the numbers will approach one trillion.

Quantum computing works different to traditional digital computers because it uses quantum states, and therefore physics as opposed to mere mathematics, as a core means for calculations and can do the same for communication. Calculations using quantum computers can break most traditionally used public key encryption methods (like factoring, discrete logs, or elliptic curves). It is becoming a critical need for most organizations to update their encryption methods to prevent adversaries with quantum computing capabilities from breaking encryption and accessing all data that previously seemed secure. This also goes for historical data.

Encryption is, of course, not enough. Also, it must be done in a form to keep our ability to analyze data so that we can improve tools, machines, and processes, and provide better services and products to all citizens. Otherwise, we would lose out on AI's potential and fall back behind China. If we get data encryption and data sharing "right," Big Data collection no longer will be an issue challenging human freedom, in the opposite. But doing what we currently do, which is allowing big governmental or corporate interests to exclusively sweep, own, and exploit humanity's data, is the definition of totalitarian surveillance.

While we are at securing our systems and data, we must therefore also put some serious firepower on this related challenge. We need mechanisms that protect data while still allowing access to it by AI. Its algorithms must be able to run effective analyses, to keep delivering the dramatic value they started contributing recently.

If a doctor looks at only a handful results of cancer tests, detecting a pattern is near impossible. If AI looks at tens of millions, though, meaningful and life-saving patters are everywhere. Blockchain-based data storage can aid in that. It documents all transactions in an encrypted and unchangeable manner, anonymized, but can also include mechanisms to access data under certain conditions.

For example, healthcare related data should remain secure and private. But AIs must be able to analyze behavioral habits, epigenetic data, biomarkers, and DNA records to find patterns, catch diseases at an early stage, and suggest preventative actions. In principle, there are ways of achieving this. Examples are homomorphic encryption, threshold proxy re-encryption, ring encryption, or blockchain tokens.

So far, though, efforts within our society focus on third parties collecting people's data. This transfers our ownership rights to them. We then let only a few private organizations control this data and commercially exploit it. And today, through public-private partnerships, law enforcement and defense organizations can screen it without stringent judicial overview. This is, in principle, the same in Western societies and in China.

Blockchains and Distributed Ledger Technologies

Blockchains and distributed ledger technologies (DLT) add to security. They, or comparable techniques, are sometimes even necessary to make sure encryption can work in practice to keep our society safe. DLT technologies provide unsurpassed fail-over and data protection capabilities. They may even survive the most extreme outside shocks, up to hacking, sabotage, EMPs, and nuclear attacks. Even if we would lose most databases, our financial and ownership records and our contracts would still exist and be safe.

DLT will therefore play a critical role in protecting our current and future computing infrastructure. It is no coincidence that encryption and DLTs underlie today's budding metaverses and virtual worlds. These are turning into cornerstones for Web3, the new Internet. I will expand on this in the final chapter.

In short, we must work on building a secure and trusted distributed infrastructure for commercial and financial transactions. Interference-free digital money using DLT will be part of that future. The question mostly is whether it will be a digitization of today's centralized fiat currencies (like the Yuan or the U.S. Dollar), Bitcoin, or a new cryptocurrency that is like Bitcoin but more democratized and better aligned with society's social priorities.

The NSA's skills and tools could play an active role in all this. And so could the technology research and development arms of the armed forces, our wider intelligence community, police, universities, and private initiatives and consortia, non-profits and research institutions.

Modern Crime Fighting

The major argument opposing encryption and distributed consensus mechanisms is that it would protect criminal and enemy activity as well. This would limit policing and prosecution powers of law enforcement (domestically) and of our military services (externally). It is a relevant concern but misses some key points.

First, many applications are heading in this direction anyway because they are more effective and efficient than alternatives. Securing systems via encryption, distributed databases with new consensus mechanisms, and quantum communication can be more economical and safer than even massive and intrusive cybersecurity systems and bureaucracies.

Second, by definition, the proposed activities would make society more secure. This makes attacks much more difficult to conduct and lowers their number. It limits illegal activities and the damage done by enemies.

Third, encryption and decentralized governance more clearly aligns motivations and goals across many groups of society. Cutting across the political spectrum, many see the surveillance and hacking capabilities and activities of the NSA, centralized governmental agencies, and private corporations as potential or actual threats to our freedoms. This would be much less so if we developed secure distributed databases, quantum resistant encryption, and so-called "self-sovereign" IDs that enable people to securely and privately conduct commercial and financial transactions on the Web3.

Fourth, the nature of blockchain or DLT technologies makes transactions not only more secure but also immutable. That term describes the fact that they document and store *all* transactions in a way that cannot be changed. This can be extremely valuable for digital forensics.

Fifth, there are alternative technical solutions for access to data and systems, in line with democratic and non-invasive principles. Encryption and access methods can be based on social groups, in some ways similar to distributed consensus mechanisms favored by blockchains. If a specified number of people within the group that establishes the consensus agrees that access to records should be granted, it is completely workable to bake this into such solutions. This is one of these areas with a significant overlap of the interests of society, government agencies, and individuals.

Sixth, tools for policing and investigation would *change*, not be eliminated. For example, extensive networks of IoT devices that constantly communicate with each other can self-detect manipulations and irregularities.

These six points show that law enforcement will not be weakened but possibly even strengthened. We just must try harder and more creatively. Top-down surveillance and complete loss of privacy do not need to remain our default answers.

Breaking up Data Monopolies - But Not Their Owners

All this will probably also trigger a transformation of the specific form of data and computing monopolies still dominating today's Internet infrastructure of the Web2. The GMAFIA (Google, Microsoft, Amazon, Facebook, IBM, and Apple) are each dominating different segments that sometimes overlap. Their unique level of ownership and control over society's data and communications must end. The simple reason is that these corporate entities' priorities are often substantially mis-aligned with the interests of society.

This is not to apply old-school anti-trust mechanisms or prevent large corporations from offering superior services and making money. In the opposite, it is to focus our attention on outcomes that benefit all parties. For this, we must properly compensate those that produce the most important ingredient of any solution. And this is *data created by individual humans.*

In "The Big Nine," Amy Webb suggests a model that in some ways is similar to what I am asking for here and could work as its starting point. It does not bedevil the corporations of America's GMAFIA itself but leverages their powers and capabilities. However, it incorporates them into a general framework and platform that is closely coordinated with social interests and aligned to the values of our society. Webb primarily aims at channeling the power of AI, which leverages many of the other technologies I mention in this book. She labels a prospective institution "GAIA" - Global Alliance on Intelligence Augmentation. This is close to what I envision. And I also agree with her premise that personal data should be secured, owned by individuals, but also widely shared across corporate and even international borders.

This is far different from what we have today. What Shoshana Zuboff calls "surveillance capitalism" is inefficient and dangerous for anybody but the Big Data corporations themselves. Above all, it currently provides motivations to maximize *advertising* by keeping us and our data and systems *deliberately insecure*. It then even *sells* that mined data to third parties. And as long as advertising and not security, privacy, or higher efficiency is the main goal being maximized, our systems will remain dangerous and inefficient for all.

The mechanisms for analyzing massive numbers of diverse data patterns have tremendous potential to improve society. But the business and revenue model is broken. It must bake in privacy, security, data ownership by individuals, and compensation for using people's data. Data is supposed to be the "new oil" of the mid-21st century. But it is not in the inanimate ground, like oil. That data originates in every single person, like blood. This makes it society's "lifeblood." As Zuboff says, sharing it should be a transaction and therefore a mutually beneficial trade, not a one-sided exploitation. There are ways of making money without expropriating all humans.

Communications

We have several major options for eliminating current weaknesses of our communications infrastructure. All of them are already in the early stages of development. This must be sped up and scaled out. The goal is to make our nationwide and even global communications infrastructure more resilient. And that would be the case, per the above, if it were diversified, securely encrypted, and decentralized without a single point of attack.

All this is possible if we would enhance today's *cellular and fiberoptic networks*, and then combine them with *peer-to-peer mesh network* communication and *satellite constellations*. Let's take these three step by step, working our way backwards.

Satellite Constellations

The first tool is to create fallback options using various high-throughput satellite constellations. The most visible example is the thousands of satellites of SpaceX Starlink. A flurry of activity is taking over this industry. As mentioned a couple of chapters ago, Euroconsult predicts for 2030 about 1,700 satellite launches per year. Most current satellite demand comes from just five constellations: OneWeb, Starlink, Guo Wang, Kuiper, and Lightspeed. In total, however, there are over 170 projects to establish constellations. This significantly expands communication options outside our atmosphere. It combines massive low-earth orbit (LEO) constellations with smaller but highly capable satellites in geo-stationary orbit (GEO).

The U.S. military is already exploring using civilian satellite communications networks, together with alternative encryption methods.[257] It would make sense to apply similar concepts to civilian communications as well. Again, there is some positive overlap between civilian and military use.

Satellite constellations reduce our dependency on transoceanic fiberoptic cables. They also add new options for communication between any two points on Earth (and potentially beyond). This clearly makes our communications infrastructure more resilient. Our goal does not even need to be complete fail-over capacity. Just being able to ensure that an attacker never can completely interrupt our critical communications would be a dramatic benefit. An adversary can take out a handful of satellites, even tens or hundreds of them. Doing the same with tens of thousands, though, is orders of magnitude more complex and expensive.

Peer-to-Peer Mesh Networks

The second way to make our communications more resilient is to establish high-performance mesh networks. These communicate peer-to-peer (P2P). Their unique characteristic is that they do not require specific hubs. Rather, the network's nodes can be as small as IoT devices. Phones, cars, watches, or glasses would work. They can talk directly to each other, forming decentralized networks. Such mesh networks are necessary to effectively use many capabilities of modern technologies.

In some ways, they work similar to how ants communicate when they are scouring their neighborhood. When they find a good food source, or a shortcut, or danger, the ants directly inform each other ("peer-to-peer"). They do not run all the way back to the nest to tell the queen who then spreads the news to all others. Going through such a central node, like the queen, is how traditional cloud computing works. The drawbacks should be obvious. It is relatively slow, inefficient, and can easily be "hacked" because it is centralized: take out or control the queen and you "own" the network.

Instead, the ants "talk" (using chemical signals) to each compatriot they encounter. These are those in their immediate vicinity that usually can best take advantage of the news anyway. They share their most critical information with others. It enables them to coordinate among themselves and spread the news rapidly to where it is most relevant. Even if several ants are "taken out," the news still would spread because there are too many ants and communication pathways. This is peer-to-peer communication in a meshed network.

It can work without the queen in the loop, without a central or controlling node that creates vulnerability. P2P works similar in the world of technology also. For example, imagine two self-driving cars encountering each other on a two-lane road. They want to warn each other about an impending collision. It would be inefficient for them to communicate by connecting to far away cloud servers. There is no need to use the bandwidth and slow down the process.

Rather, they can talk directly to each other using vehicle-to-vehicle (V2V) communication like dedicated short-range communications (DSRC). If all cars on the road would do that, they could also use the same technology to relay information between other cars or other devices. They would become part of a dynamic and extremely resilient mesh of communication nodes. Without cables, cell phone towers, or satellites, all cars could communicate and transmit data even to far-away locations. The same mechanism can also work with all other devices of the Internet of Things. And as hundreds of billions of devices blanket our planet, this enables the just-described solution.

Such a setup would then not require dedicated intermediaries like cell phone towers or fiberoptic cables. This means we have more options for communication. The potential pathways even increase exponentially. Communications can use hundreds of thousands of cell towers, hundreds of millions of vehicles, hundreds of billions of IoT devices. Even the most powerful, most advanced, and most dedicated attacker would find it impossible to cause a complete breakdown of communications.

This is the way the world is likely heading anyway. Our government agencies should encourage and support the trend, hopefully speeding it up. Military devices and civilian ones can communicate resiliently. The benefits go far beyond the coordination of self-driving cars and drones.

Quantum Communication

The third way to secure communications is via quantum communications. This is not just about quantum computer's ability to break traditional public key encryption. Rather, it uses quantum physics to exchange information or secure the process.

We may only be a handful of years away from the most frequently used traditional encryption methods being breakable by quantum computers. Maybe it will happen even faster. A task comparable to the Y2K challenge may have to get under way pretty soon across all industries. It would analyze and upgrade all essential data protection and encryption tasks within our societies.

Besides *breaking* encryption, though, quantum computing can also *secure* communications. Quantum communication securely exchanges information by linking photons at locations far from each other. It uses a scientific principle called "quantum entanglement." This is un-hackable because, the way quantum physics works, the information will automatically change once observed.

Quantum communication already enables secure remote communication even over large distances. For example, already in the mid-2010s, Chinese scientists could use such a quantum mechanism to send large files between a satellite and a ground station. And to distribute quantum keys securely, quantum effects and photons can make sure that an adversary cannot listen in. To test this, China created, in 2021, an extensive quantum communications network.[258]

Quantum key distribution (QKD) is another way of securing communication. It uses some of the same quantum aspects of physics, but mainly to protect today's encryption methods when sending encryption keys via fiberoptic cables. QKD already is being used, for example in the financial industry.[259]

Electricity & Energy Security

The world is encountering serious issues and bottlenecks during the current audacious "energy transition." Its goal is to use environmentally friendly sources of energy and electrical power in all aspects of our lives. Critical challenges remain. Among them are the intermittency of many renewable sources of energy, and the need to extend the capacity of our electricity infrastructure by adding vast numbers of charging stations, substations, and high-voltage lines. In the meantime, as pointed out in Chapter 12 above, the size and operating principles of today's electrical power grids make them greatly vulnerable.

To overcome this, we must decentralize and localize, establishing smaller and potentially autonomous microgrids or even nanogrids. These terms describe the ability of cities, communities, military installations, and even neighborhoods or office parks to decouple from other grids and operate independently. Like with the communications network, resilience does not have to mean complete fail-over capacity. A home, or even a city, does not need to produce all of its energy locally, as desirable as this may be for other reasons. Rather, it is sufficient to cover the most basic requirements for the respective community in case of an attack or other emergency.

Smart mechanisms governed by blockchain mechanisms would enable load balancing. They could share stored energy and reserves and optimize electricity usage efficiently. Without a central point of control, attacks would be much less likely to succeed.

We can achieve this, as I outlined before, with locally available renewable sources of energy like solar, wind, or geothermal, grid-level storage of electricity, and large home batteries. And all this can be for a city, town, street, building, or even individual house. Again, we don't need perfection. The goal here is to make it impossible for an attacker to bring down the grid and pipelines and deny most people within an entire society the absolute essentials to live. Such essentials include the use of stoves, fridges, freezers, automobiles, trucks, and digital devices.

It does not have to end there, though. Generators using fossil fuels or hydrogen can add more resilience. Modern designs of nuclear power could even become the backbone of most small grids. For environmental reasons, and probably for cost reasons also, we must consider a massive expansion of such modular nuclear reactors. Even space-based solar power (SBSP) may be an alternative.[260] This is unlikely within the very near term, although there already are several European SBSP initiatives.

We can even go one step further. Part of the challenge with today's system is its need to transmit electricity from power plants to end users. Locally produced energy would cut down the need for high-voltage lines. As pointed out in Chapter 12, these currently are a relatively easy target for attacks. There are alternative solutions for the current transmission network, though, and more flexible ones.

Similar to how IoT or V2V mesh networks can provide highly resilient decentralized and dynamic communications, the transmission of electricity can use the same kinds of principles. Here, the P2P concept is called vehicle-to-grid (V2G). Its underlying premise is that any e-vehicle could in theory function not just as a user of electricity but also as a transmission device that delivers energy. For example, houses or neighborhoods can collect solar or geothermal energy during the day and store it in a large battery. When the car owner comes home from work, the car connects to that battery and charges up. The next day, at work, this car battery can charge batteries at the office.

The result of the above measures would use economics to turn the tables on attackers. Today, it may take no more than bringing down a few dozen substations, cutting multiple high-voltage power lines, and blowing up a few pipelines to bring most of the country to its knees. The system I just described would require the attack of every microgrid and nanogrid across the whole country to trigger a similar scale of results. And this would still leave most sources of energy generation untouched, and many distribution mechanisms would or could still work (like cars as transmission devices, or hydrogen).

As part of our transition to renewable sources of energy, we have already started with some of the above. As a matter of national security, this must continue and speed up. Accelerating the energy transition may not even require specific governmental direction or management, though. In the opposite, it would probably even be quicker and more efficient without either (particularly without governmental *management*). Or how many localities would prefer less resilience, higher infrastructure costs, and increased dependency on others far away? All it may take is remove regulatory limits and facilitate industrial standards for charging and transmission.

Supply Chains

In Chapter 12, I described how the breaking down of energy and communications infrastructure can prevent nearly any products, including food, from reaching our stores. The just mentioned measures would significantly reduce the likelihood of this happening. Our supply chains will be much more resilient if energy, computing, and communications stay available. This ensures our ability to transport products, cool food, operate warehouses, and run businesses. That alone eliminates most attack points.

Supply chains also must secure shipments of raw materials. And then, we also have to secure the main value component of modern products, whose nature is digital. This brings us back to encryption and to using blockchains as distributed, shared, and un-hackable databases. We must protect the ownership of data and digitized knowledge in this emerging new world of the Web3.

There is more we can do, though. We can take extra steps in securing critical supplies. Both take advantage of digitization, although in a different way.

Near-Shoring: Producing Locally

The first one is to *produce locally*. This is also called "near-shoring." We can do this if we maximize digital and capital contributions instead of labor. This enhances the flexibility of businesses. Near-shoring manufacturing is not about creating jobs in factories. This may be a welcome side effect. But mainly, robotics will take over most traditional blue-collar work and therefore change job profiles. The new supply chains are predominantly digital. Most human design and manufacturing work will be conducted online. It will happen via virtual worlds that use digital twins. The physical assembly of new products will therefore require less shipping of semi-finished or finished products across vast oceans.

This is consistent with the continuous shift toward digitized design, capital intense manufacturing leveraging robotics and AI, and the lowering of labor's contribution to the costs of products. An example is Tesla, which produces "American-made" cars in California.

According to an analysis by Eric Kosak of CleanTechnica, Tesla's total labor costs are about $3,000 per car, or roughly 5% of each car's sales price.[261] Taxes and shipping costs are likely to have a bigger impact on the car's price than labor. The largest individual portion is not capital costs either. It is the cost of materials.

But this is only for the final stages of assembly. And only some of these materials are raw materials. The majority are semi-finished products or components (like batteries). Electrical vehicles require fewer moving parts than combustion engines. Even so, each one must be designed, developed, and fabricated as well. Kosak goes through several iterations of identifying the major cost buckets during each step along the supply chain. He does this until he reaches the point right after raw materials have been pulled out of the ground and purified.

As a result, we can make an educated guess that the biggest aggregated cost components of a Tesla car are R&D and other digital intangibles. They make up between 30% to 50% of the total cost along the supply chain. Raw materials are above 25%, and the rest are capital plus labor-related costs, roughly evenly split. The main point here is that, above all, digital contributions, and then raw materials, make up the overwhelming part of a car's value. And we can expect the share of both to keep increasing. This is digitization in a nutshell. It also is likely typical for many, if not most, other complex products.

In theory, any business in an advanced economy can construct a highly automated factory. It will require mostly capital for robots and the digitized know-how to operate them. The digital know-how can be generated anywhere and moved at the speed of light. With the current advances in augmented reality, such factories can even mostly be operated and maintained remotely as well. Therefore, global supply chains can and will shift, moving fabrication into highly digitized and capitalized societies. The main input will be, as in the above Tesla example, digital intangibles and raw materials, then capital costs. If we take advantage of this, the just described process will significantly benefit the United States and other Western countries.

Should we then source most of these raw materials domestically or from allies, we just would have reduced significantly our need to protect physical supply chains. They have either moved into the digital world or are physically close to us. Even where physical supply chains or subcomponents still would exist, most would have shifted away from traditional shipping routes.

New technologies will intensify this effect. Singapore has recently been successful in taking advantage of the trend toward robotic and additive manufacturing.[262] In general, the times of traditional mass production are starting to phase out anyway. Modern products are increasingly customized and produced in smaller batches. This can be done at any location that has access to raw materials and machinery. For about a decade already, such locations include aircraft carriers.[263]

Food Security and City Agriculture

The second step involves our *food security*. Some similar principles apply to agriculture as to manufacturing. Currently, the world, including the United States, highly depends on fossil fuels as input for producing enough food. This means that we must find ways of reducing the need for fertilizers or replacing the Haber-Bosch process. Otherwise, we cannot feed the world population.

An important part of the solution will be new agricultural methods like vertical farms to limit our dependency on supplies from far away. Today, this is possible for greens and vegetables. Locally grown food already is a reality, cost-efficiently and environmentally friendly produced in inner-city warehouses and skyscrapers. We can also lab-grow amino-acids in form of fish and artificial or 3D-printed meats. Hydroponics and vertical farming can contribute to a consistent locally produced supply of high-quality vegetables and potentially even fish or meat-like nutrition.

It will be very difficult to completely replace the use of the large land areas for farming that currently provide most of our grain-based foods. There also is a continuing debate about whether plant-based or meat-based nutrition is healthier and more beneficial to the environment (i.e., considering top soil maintenance and regeneration vs. greenhouse emissions) and particularly for wild animals (i.e. the impact of large-area farming on animal habitats and biodiversity).

However, emerging technologies are likely to help. Today's most advanced forms of precision agriculture leverage satellite and drone-based monitoring of fields, sometimes down to individual plants. AI algorithms can control autonomous drones to optimize the use of fertilizers and water. When this additionally and sensibly takes advantage of more resilient genetically modified organisms (GMOs), we can make huge strides to bring food production closer to consumers.

Much of what I just described used technical terms and involved topics far from the killing machines that come to mind when thinking about the military and war. And yet, these measures will tremendously strengthen our defense. Not only will our military be less vulnerable, more capable, and better focused. Our society will also be much more resilient. The organization tasked with militarily protecting Americans is called "Department of *Defense*." All I mentioned so far is effective defense at its core.

This context means that military and civilian solutions are reinforcing each other. In shifting our focus to defending our infrastructure, we will have improved our society as well.

Once we achieve the above, we have effectively used modern technology to turn the economics on prospective attackers, including assaults from the Chinese PLA during a war. The asymmetric advantage will favor our defense. We will be resilient. It will be unlikely that large patches of the country completely lose access to energy or communication or food and critical products. To completely bring down our society would require something akin to all-out nuclear war or a massive attack using biological weapons.

China can do the same, which is fine and fair. But now it is time to take the last step, and the most important one. It is one that China cannot match, at least not unless it changes its political system and approach to organize its society.

And therefore, let us embrace systemic competition and confidently go into the offense. Let's find out whose system really is superior.

15

A Liberal World

Offense: Protecting and Shaping the Web3 and Digital Metaspaces

A baby is born. Like about 111,000,000,000 before it. This one, though, is different.

The story I am telling here is real. It hasn't happened yet, but it definitely will, soon.

108 billion was the estimated total number of humans, of homo sapiens, born on Earth by the year 2020.[264] On a large scale, birth is a common event. In each specific case and particularly for its parents, however, the birth of a baby is always something truly special.

The specific baby we are talking about here is unique to all of humankind, though. Just around the time when the above total number of humans ever born hits 111 billion, maybe a bit sooner or later, this baby's birth will be special compared to all other humans. Such a birth will for sure never have happened before, ever.

And it will not add to the above number.

This new baby will see the light of the world not on Earth. Instead, it will be born in space, on the moon, maybe even on Mars. For the first time in human history.

Until a few years ago, this would have been impossible. Today, even though we still have some "kinks" to work out,[265] there is nothing technically preventing it from happening. We just don't have (yet) a reason. In a couple of decades, the mere existence of the first permanent off-Earth settlements may be reason enough. The baby then even may be born, grow up, and never step foot on planet Earth throughout her entire life.

This little tale shows how profoundly the world is changing today. Looking beyond traditional borders and even beyond our planet changes the meaning of the term "geo"-politics in international affairs. Right now, we are laying the foundation for this future, which will increasingly become important within the 2030s and 2040s.

But until then, and at least for the next ten years, the most profound *and noticeable* technological disruption will not yet come from space. It will be *the digitization of nearly everything* and the emerging metaworlds of the *Web3*. They promise to tear down all traditional barriers between organizations.

All this has a direct impact on the question of war and peace. It affects strategies and institutions for keeping us free, prosperous, and peaceful. Domestically, whichever society most effectively uses these upheavals and opportunities can deliver the best result for its people. If it succeeds, it then will move ahead, in hard power where relevant, and in the soft power appeal that complements it. Success and failure, though, will depend on how this digitization will be designed and governed. Above all, will it be free or authoritarian?

This final chapter addresses the third set of steps toward a peaceful and prosperous future. It urges accepting China's challenge and to compete with the PRC's system. I ask from us to prove that a liberal system indeed is more effective and better for the people. Not just at home, but also internationally, we can and must set a counterpoint to China's top-down controlled and managed world.

The key for this is *using 21st century technologies to develop a rules-based liberal system for the 21st century*. Our focus must be on where the near future is heading, which is the emerging digital platforms of an Internet that is turning into the Web3. Within the Web3, we must safeguard the identities of people and alternate forms of legal entities. Then we must secure their data and communications, extend legal property rights to this data and to digital products, and facilitate and protect their personal and commercial activities.

It is this, or us continuing to follow China's lead toward de facto totalitarian-style surveillance and technocratic manipulation and control of the population according to the utilitarian vision of small groups acting as "elites." We cannot have both top-down surveillance and controls and bottom-up freedom and market forces.

Accepting the Challenge

China has not figured out how to win in this game either. However, it seems willing to do whatever it takes to get there quickly. The PRC has no qualms about disrupting its society whenever it sees obstacles or opportunities in leveraging technology to improve it. If we really believe in the superiority of our own system and society, then we should do the same, although our way.[266]

China's meteoric rise during the past four decades has been a dramatic benefit to its people and many others in the world. As its growth slowed down in recent years, the world has become awash with smart and often correct explanations of Chinese weaknesses. These mention debt, shrinking population numbers, aging, lack of diversity, waste, corruption, and much more. But little of these matter unless we get our own house in *better* order than China does. As long as China grows faster than us and continues to more pragmatically and efficiently deliver value to its citizens, the most important part of the trajectory remains unchanged.

For why would the rest of the world trust our system and capabilities, and follow our lead, if they can see that their people may be better off following a different pathway? Wouldn't they also want to champion interventionist policies and tolerate surveillance and limitations of individual rights, if they believe that these are the price or even condition for prosperity and security? This keeps being advocated, and sometimes done, in many developing countries and even in the United States. It has become fashionable for politicians to associate China's success with the belief that elites know better, and top-down actions and directions create better results.[267]

Digitization Trumps War

The impact of the transformative consequences of today's technology-driven upheavals is orders of magnitudes larger than gaining control over an island. It goes far beyond almost everything commonly seen as causes for war, even in East and Southeast Asia. Whether an authoritarian elite or democratic institutions (and elites) govern Taiwan or, for that matter even Turkey or Colombia, pales compared to how the world deals with exponentially growing digital technologies.

When we process the scale and profound nature of the shifts that are coming at us, we simply cannot afford to lose our focus and risk near-everything in a fight over a hill, or an island. Even a large and wealthy island with millions of people.

However, we only can win a decades-long struggle among systems if we treat it as such. In the past two chapters, I covered how our military must adapt, and how to re-redesign core infrastructure in our homeland to make it more resilient. But it is most important for our long-term success that we also transform the way we interact with our people and with the world. We must design the larger political and economic system for what is coming.

This is because today's systems mostly use outdated concepts envisioned and originating in the late 19th century, developed in the early 20th century, and then optimized in the 20th century's second half. Their design reflects the challenges and capabilities of those times in response to the industrial revolution. It does not consider the ones of the mid-21st century and the demands and opportunities of the data-driven digital revolution powered by AI. Our systems, frequently, still focus on economies of scale via massive and often hierarchically controlled organizations, whether these are governments or private organizations.

Such centralized institutions have facilitated many of our greatest advances. Increasingly, though, they keep falling behind the capabilities of new decentralized solutions. Today's most advanced technical systems and organizational structures are distributed and self-organizing, powered by individual nodes and empowering them. They shift decision-making "down," often all the way to individuals. Examples are personalization from clothes to healthcare, instantaneous communication, access to information and knowledge via smartphones, biohacking using wearables and health tests, distributed consensus mechanisms of blockchains, and edge computing. Some call this "democratization" of technological capabilities.[268]

This not only sounds much more like liberal democracy and market economy rather than technocratic totalitarianism. It also is exactly that.

Pursuit of Happiness - or of A Dumber, Poorer, Weaker China?

If we were to fight a war against China, though, or any other potential challenger on the Eurasian landmass, what would we want to achieve? Is it preservation of a status quo no matter what? Or regime change in China, making sure its society's domestic priorities and system are the same as ours? Or would it be to prevent that China, internationally, becomes more powerful than us and our allies? And if so, how far would we be willing to go to ensure success?

These are very important questions. The only other way for the U.S. to stay on top of the food chain would be for us, and for the larger system represented by America and its allies, to be overall more successful economically. Unless we grow stronger more rapidly than the PRC, though, China will become stronger, absolutely and relative to us.

It will continue to achieve this in three steps. First, China will keep accumulating scientific and technological knowledge and competence. Second, it will translate these technological capabilities into economic power and solutions for its society's challenges. And third, it will then turn these technological and economic advances into military power.

We can stay ahead only by being better at this - or preventing China from achieving at least one of these. It means that we would have to keep China dumber, poorer, or weaker than ourselves. How is that for ethical superiority?

But this is where negative and zero-sum thinking gets us. It directly contradicts the past several hundred years' trajectory of almost all meaningful factors of human progress observable. Because innovation has assured that the world is not static and zero-sum. Rather, it keeps improving by almost any significant measure.

Many thinkers, academics, and doers have pointed this out. Among their diverse group are Steven Pinker, Matt Ridley, Hans Rosling, Cathie Wood, Peter Diamandis, Thomas Sowell, Bill Gates, Deirdre McCloskey, and Peter Thiel. Equally notable, it directly clashes with how Singaporean Kishore Mahbubani so eloquently keeps depicting the long arc of history on which Asia's recent re-emergence is built on.[269]

Indeed, a lot of evidence suggests that today's most healthy, most wealthy, and most educated times in history, ever, can continue to make even more progress that benefits all. We can translate this into a positive global strategy. It should be easier, and make more sense, for us trying to *stay above* China rather than keeping it below us. It would then not even matter much whether we succeed or fail relative to others, even to China. As long as we advance, our people would be better off. They would win either way.

This is not simple idealism or wishful thinking. Rather, it is an essential component for advancing humanity while preventing great power war and global disaster. Fact is that the techtonic plates of the world are changing. I described them in Chapters 3 to 7 above. And this has a transformative impact on geopolitics and power relations inside and between societies.

By adhering to old rules and institutions, we are therefore fighting a losing battle. This inherently weakens us, and eventually all those aging and creaking systems we hold so dear will break anyway. And that would be catastrophic for us and for humanity. Because: who and what would be left, who and what would "win?"

A Liberal - or "Capitalist"? - Approach

Let's contrast this with the alternative, and we can see opportunities everywhere and embrace or even seek them. This vision wants others also to be successful and become wealthier, including China. Success enlarges markets and drives innovation. It increases all our opportunities to live even better lives and be more prosperous. This keeps enabling new and abundant possibilities for our societies and each individual human. Such empowering of each person makes liberal systems a better fit for this new world than authoritarian alternatives.

It is near impossible for any human to entirely comprehend and process the pace of 21st century technological transformations. And this turns this chapter here into the most important part of the book. It outlines the approach for addressing the most critical challenges confronting us because of exponential technologies.

During times of rapid change, it is necessary to focus on opportunities instead of threats to succeed with our most important responsibility. And that is, as Spanish economist Daniel Lacalle says, "to maximize the level of prosperity, access to education, and the generation of opportunities for all."[270]

The result of a job well done is a continuously more abundant world with extraordinary, rapidly accelerating opportunities rooted in ever larger productivity. This growth of opportunity is "exponential," at least currently. The reason is simple. Many of its underlying technologies are growing at exponential rates and even in a mutually re-enforcing manner.

In his seminal 2001 article of the same title, renowned inventor, computer scientist, and futurist Ray Kurzweil calls this "The Law of Accelerating Returns."[271] The two decades after he published his article have so far proved him right. We subsequently have seen dramatic advances in robotics, genomics, communications, alternative energies, battery technologies, biotechnologies, battling of diseases, and space technologies. They enabled profoundly new business models. Almost all of this was driven by the exponentially growing digitization of our world that Kurzweil forecasted.

Nobody can contain this. And I sure don't *want* to contain it. Stifling it would directly contradict universal aspirations of humanity. And writing as an American, it would violate the ideas and ideals reflected in America's Declaration of Independence.

Everybody can take advantage of this. All we must achieve is organize our societies in a way that empowers and motivates everybody. When we maximize opportunities for all, we will have a great ride.

Easy said. So how do we do it?

Toward a Modern Rules-Based Liberal System

There are several crucial conditions for establishing a modern rules-based liberal system, mainly four of them. These should be supportable by all democratic and liberal societies.

A. **First** is global and distributed connectivity for all people. We must "liberate" and protect space-based satellite constellations and ground-based networks and integrate them in a way that ensures effective and secure global communications for all. This will not only *allow* people to explore, innovate, and speak freely but also *enable* them to do so.

B. **Second**, we must secure data and communication to not just establish digital ownership *rights* but also effective *factual ownership*. This is not just the esoteric question as which it so often seems to be treated. It centers on private property. In a digitized world and in Web3, only tools like encryption and distributed ledgers can ensure property rights to the most important assets of the 21st century: data, whether generated by people or by the machines serving them. And then, this must also enable their effective analysis using AI, while preserving privacy. Our collective data is an unprecedented treasure trove. Enabling artificial intelligence to analyze it enables us to reap its benefits. Where China can use data from 1.4 billion people, Web3 can reach billions more if it treats our data and the digital world as "commons" and not as private property of a handful of businesses.

C. **Third**, new platforms and infrastructure must protect identities of people and organizations. Most notably, this creates the ability to share and communicate via real identities, digital avatars, and digital twins. In Web3, people and organizations will do this in virtual worlds that are connected to the physical.

D. **Fourth**, Web3 must contain the ability for anybody to conduct commercial transactions and to settle them online. This must include new systems to transfer value, including money. Invariably, this will require using a scalable digital currency managed in distributed fashion to settle financial transactions without interference and meddling.

All four conditions together effectively and powerfully define Web3 as a decentralized liberal system. It then will empower individuals and facilitate innovation and commerce.

The consumer-oriented Internet of the 1990s and early 2000s, or Web1, was about reading. Everybody started using email and websites to share information. The one of the 2000s and of the 2010s, Web2, was about contributing. This made it easy to share data. Social media turned everybody into "contributors."

In the 2020s, we now have started building the Web3, the new infrastructure that I keep talking about. This process will probably continue for at least the next couple of decades. Web3 is about ownership and commerce. Nearly all commerce will happen in the digital world. If we want to buy or sell physical products or services, we must therefore integrate them into these digital worlds. The tools for this are called "digital twins."

This differs from what we today often call "eCommerce." That concept, as I described above, merely facilitates traditional commerce with electronic tools. We receive invoices per emails instead of the postal service, go online instead of calling the vendor to find out where in transit our products currently are, and so on.

"Real" digital commerce, though, *creates* the core value in the digital space, and keeps it there. This happens by cooperatively designing, showcasing products, creating content, executing contracts, generating data, and programming code within the virtual commons themselves. Fabrication, then, merely involves sending the agreed upon and finished design to an automated fabrication facility. There, robots create the physical representation of the digital design. The "eCommerce" part of shipping a product is just a tiny part of this process.

My first point, about "liberating" and securing communications, warrants a few more words. Today, the electromagnetic spectrum is near-completely controlled by nations. This enables them to censure, manipulate, and control their people's communication, increasingly turning the Internet into a "Splinternet." The rise of satellite constellations challenges this dominance. It creates a unique opportunity because satellites are physically removed from geographical limitations. In theory, anybody with a satellite dish can technically link to Starlink or other constellations. Limitations are artificially introduced through national laws and bureaucracies. This leaves power with politicians, often with questionable legitimacy, controlling motivations, and authoritarian streaks.

The U.S. should aggressively advocate to free up a portion of the global radio frequency spectrum for such global communication networks. It then should put this part of the spectrum under governance of a non-governmental institution jointly organized by a community. We could loosely model this body on the 1990s role of the Internic that governed the Internet.

Again, the goal is to not only allow free speech but to enable it. Once everybody is technically able to communicate and conduct commerce, globally, independent of governments or large corporations, we will have liberated the Internet. This time, though, it is not just the Web1 or Web2 but the much more important Web3. The effect on innovation and wealth-generation will be dramatically positive.

It will expand individual education and global business opportunities for locals anywhere. In a limited although slightly unique form, we have seen this working in Ukraine against an invading force when SpaceX sent satellite dishes that could connect to Starlink. A free Web3 will ensure that any person in Kenya, Canada, Bangladesh, Spain, Texas, or Peru only needs a computer and a connection to this satellite network to access information, communicate, participate in the global economy and, therefore, to prosper. Neither governments nor corporations could prevent this from happening.

Modern and Liberal 21st Century Domestic Solutions for Society

Future-oriented 21st century solutions must include 21st century thinking. They must establish new infrastructure and systems that trust the people, empower them, and tap into the near-unlimited innovative power of individuals no matter where they are physically located.

The alternative would be to do it China's way. In that case, the government, controlled by an ideological and political group (in the PRC this is the Chinese Communist Party), would wield all the power. It would use complete surveillance and empower AIs to mold and engineer a society according to a central vision. This would by definition be elitist. The government would control all identities, surveil all activities, have AIs enforce laws and rules, and optimize the system according to *its* priorities and *its* interpretation of vaguely stated principles and goals. It eliminates political debate and competition.

This neither works for domestic nor international cooperation in a free and liberal world. There, the rules must be the same for all. Power comes from the bottom. It is messy. It involves free speech and debates and pushing the envelope. It believes that all humans have their own unique set of priorities and definition of "happiness" and should not be molded into a narrowly defined common one. People are not just robotic pawns in a utilitarian sense, being guided from the top through positive and negative reinforcement to conform to a uniform worldview.

A New Rules-Based Liberal System

The most significant challenge of our time is to create such a bottom-up distributed rules-based technical infrastructure, as the baseline for *a new liberal system*. It must differ from old ones because technological capabilities transform and transcend traditional organizational structures.

For example, within a handful of years, hundreds of billions and soon after even trillions of sensors of the Internet of Things will collect massive amounts of data and make them available for AIs to analyze and optimize in near-real-time. Traditional human decision-making institutions and processes are orders of magnitude too small to create efficient solutions. "Bureaucracies" cannot function at all anymore because humans are too slow. They are incapable of processing this deluge of data. AI-enforced systems can better optimize traffic, keep humans healthier at lower costs, and enforce laws more consistently. But we must make sure these systems' design is in line with our principles.

If we want to win the future, protecting specific mechanisms of old institutions designed in and for a completely different time must not be a priority. This primarily applies to outdated bureaucratic systems in which humans fill out and move paper or electronic files. We cannot afford to tinker on the edges. An example for tinkering is to automate file transfers or share data. Such actions do not get to the essence of not needing such files in the first place. If *everything* is being recorded and stored, and accessible anyway, what would be the purpose? Rather, we must face the core challenge and leverage the revolutionary technological capabilities that can move us from the 19th into the middle of the 21st century. Sticking the head in the sand is not a strategy.

When competing with China's solutions to its society's challenges and opportunities, we must re-envision many of our own as well. This means that the specific functioning of all institutions must be on the table if this is necessary to make our society and our people better off. There are plenty of aspects that could be transformed profoundly. For example, many aspects of our current healthcare, education, energy, pension, finance, and even legal systems are becoming either dysfunctional, unsustainable, or both.

Rather than trying to squeeze marginal value out of these, we must not just adapt them, but change the underlying ways of how we organize many parts of our society. Like the industrial revolution changed the role of government and corporations, the digital revolution must do the same.

I will just point out two examples.

Health Care

It makes little sense to continue spending outrageous amounts of money on "sick-care" using 19th century and early 20th century-style "health" insurances. We should stop such inefficient practices that often do little more than protecting bureaucracies within governments and the private sector. Too much, they seem about jobs, and profits for large businesses, rather than delivering high-quality outcomes to people. Today, "real" healthcare is all about data-driven personalizing, predicting, and *preventing* sickness. Ever more sensors and data are moving into the hands of individuals.

Currently, hospitals and clinics make money by treating diseases. Public and private health insurances make money by being middlemen that take money from people and send it to hospitals and other care providers to pay for said treatments. In the U.S., very roughly expressed, these two functions generate revenues of about one tenth of the economy, far over one trillion dollars. [272] Therefore, preventing diseases challenges all such established interests and structures. But we can also see the size of the opportunity, not even to speak of the economic boon when tens of millions of sick or old stay productive in society instead of using resources merely to survive.

The traditional sick-care system cannot cope with this and must be near-completely overhauled, transformed, and probably even transcended. How to do that, specifically, is our challenge. China and others will have their answers. Let's make ours as good or even better.

What Means "Rules-Based" Anyway? - The Legal System

Most scary to many in the West, we also must address the major shortcomings of our legal system that technology keeps revealing. Way too often, people ignore the reality of terms like "rules-based" and "rule of law."

Today, our legal system consists of vast numbers of often contradictory laws and regulations that we usually enforce only selectively, most of them even *rarely*. This cannot continue in a world where *everything is monitored and recorded*. It must cope with the fact that hundreds of billions of IoT sensors will document and know *any* human action, with most of these actions being guided and managed by autonomous and semi-autonomous systems. This leads to a serious dilemma. Today, everybody violates some laws or regulations every day, and if it is just exceeding the speed limit or not completely stopping at a stop sign. It probably includes also the at least "three felonies a day" that the average American commits per the 2011 book written by Harvey Silverglate and Alan Dershowitz.[273]

The real issue is that all these violations are *documented*, and therefore, *can* be prosecuted. Will they, though? And how? The current legal system can neither deal with billions of violations each week, and neither is it possible or workable to throw every person into jail. Saying "nobody is above the law" sounds good. However, today's reality is exactly the opposite because in many areas *everybody* seems above the law.

We are - temporarily - moving far from the question of war and peace over Taiwan. But there is a point because this addresses the nature of our own system as opposed to China's. So, let us contemplate for a moment what it means that every traffic infraction, every single act of misbehavior, and all surrounding hard evidence will be documented by our systems, and *can* be known almost instantaneously by AIs. This is already the case, or is going to be so soon, in many parts of both China and the U.S. How will governments use this information? Which society will establish systems with fairer and more effective outcomes, and how?

Due process, traditional juries and court procedures, lawyers sifting through hundreds or thousands of often contradictory codes of law, and even traditional forms of sentencing - none of this can continue to work as it did in the past. The system would be overwhelmed by hundreds of millions of cases - each day! What, then, is the answer? Dramatically cutting the numbers of laws?

Introducing a governmental "social credit system" of some kind, instead of sentencing? And therefore, circling back to China's solution, decried by many in the West as totalitarian? Or do we keep ignoring almost all violations and apply the law only selectively, as we have been doing for many decades already? What would this contradiction leave of our cherished system based on the "rule of law?"

We must find honest and *effective* solutions for such dilemmas. The world is watching. Will China's be better at it than us?

Potential Solutions

This goes beyond healthcare and the legal system, of course.

In the previous chapter, I outlined how the creation of microgrids, owned and operated in distributed fashion by small communities, adds resilience and efficiencies to the energy sector. By removing them from centralized control, it also can align their goals to deliver energy with the goals of the community they serve. They can do so in an efficient, resilient, and environmentally friendly way.

We will similarly have to re-consider and transform not just how we organize healthcare (e.g., introduce preventative predictive and personalized healthcare via token systems), but also education (e.g., link educational value to income or value generated in the future), transportation (e.g., establish P2P communication and self-governing systems for autonomous driving and flying), and much more. As I just described, traditional top-down bureaucracies are incapable of keeping them function effectively in the data- and sensor-rich mid-21st century. We must move them out of slow-moving and political control into more focused and all-inclusive markets that are closer to the communities they serve.

For example, like other products and services, healthcare solutions also get more efficient the more data they have. Unlike with current health insurances, this motivates inclusiveness. Everybody can become healthier, and make everybody else (on average) healthier, simply by participating. Let us just make sure that "everybody" also shares ownership - to their data and to the system's outcome. In tech-speak, we may think of it as socializing critical human-centered systems by introducing market incentives to optimize the value of a public good.

But the details are not as important as *that* we solve these challenges, using modern technologies and maximizing value for all our citizens, and in line with our principles. If we achieve this, we "win" in our competition with Chinese-style authoritarian systems. If we continue to compromise our values by centralizing, bureaucratizing, surveilling, and top-down controlling via inefficient bureaucracies dominated by narrow elites and special interests, then we cannot win. Then China has won because we have mimicked its society and system. And what would then be the point of strongly opposing it?

I am optimistic about the superiority of liberal and market-oriented solutions that are democratic, self-organizing, and distributed. Systems of the industrial revolution, whether monopolist capitalist or communist cannot solve today's challenges as effectively. Simply put, centralized organizations cannot compete with the efficiencies generated by self-organizing social groups - like democracies and markets. Modern technology can create much more specific and therefore efficient value than ever before. It does so through *more* liberal organization, not less.

A 21st Century Liberal World: Revisiting Alliances

There is no reason why these solutions would have to be limited to a country's borders. Rather, such technical platforms and frameworks can become the core of a modern, genuinely liberal, globally available system delivering superior results. Focusing on the digital world as the world's vital engine of innovation and commerce, we can develop a global rules-based liberal system for the future. It comprises the current old Internet, Web3 platforms, metaworlds, and their extensions into space.

These are the most crucial areas in which the contest among superpowers will take place, not land or shipping lanes. We cannot settle a 21st century rivalry and competition with missiles, fighter planes, tanks, submarines, or aircraft carriers. We must do it via innovation and by using technologies such as AI, data analytics, quantum computing, blockchain, AR and VR, satellite constellations, 6G, and the Internet of Things. The outcome of U.S.-Chinese competition will be determined by how we structure and organize these commons of the digital world and of space.

This must reflect the reality of how profoundly the international world changed compared to the second half of the 20th century. As a consequence, our alliances and international institutions must adapt also.

At the outset of his "The Art of War in an Age of Peace," Michael O'Hanlon lists America's "security partners." He concludes that about 75% of the world's GDP is being generated by countries that have at least some level of partnership with the U.S.[274] According to this list, countries overall producing almost 70% of global GDP subscribe to the principle of a rules-based liberal global system as well.

O'Hanlon's list is helpful although slightly skewed in favor of the U.S. because he measures GDP by its nominal value and not by Purchasing Power Parity. PPP is how the CIA and the World Bank compute economic output. No matter the measurement, though, it contains the just-mentioned core Western nations with roughly one billion people in North America, Europe, Japan, South Korea, Australia, and some of their neighbors.

We must revisit and adjust the nature and institutions of these alliances. Current formal alliances include the treaty allies of NATO, Japan, South Korea, the Philippines, Australia, and New Zealand. They extend to the tight bonds the United States has particularly with Israel, Singapore, Thailand, Malaysia, India, Switzerland, Brazil, and many other countries in the rest of the world. Whether or not formally allied, all favor at least nominally a democratic society and free market-oriented economies. They are part of what many describe as the "rules-based liberal world order."

These current alliance systems, though, focus on the military. They have a hub-and-spoke structure with the U.S. at the center, connected to various geographic world regions. European countries and Canada are allied with the U.S. to protect their "North Atlantic" interests and nations. In East Asia, Japan and South Korea have treaties with America for the equivalent reason. And so do Australia and New Zealand in the Southern Pacific. At least nominally, the Rio Pact achieves the same for almost all of Latin America.

This works well enough for geographically limited conflicts. Even then, it spreads the United States military thin when facing any military challenge on another continent. But the situation gets even worse when confronting another great power. A hub-and-spoke model does not work well for a global *systemic* struggle. It defines this struggle not as a conflict or competition between systems but between two nations, the U.S. and China. The disproportionate burden remains with the U.S. even though it only contributes roughly one third of the overall population and half the economic power.

Compared to when these alliances were established three generations ago, we have profoundly new and different threats to our freedom in the globalized and digitized world of the mid-21st century. This also introduces novel strategic options for defense. The interests of countries as diverse as Japan, Singapore, Australia, Canada, Germany, Sweden, Estonia, and the United States are more closely aligned than ever before. And the main ways to generate, maintain, and use power lies in technologies that make borders and geography matter ever less. They keep shifting into the digital and space.

We can therefore make sure that we institutionalize our friends' common interests independent of their geographical location. Indeed, in recent years, we have seen the tentative emergence of such a multi-continental alliance, informally via the Quad of Japan, Australia, India, and the U.S. and formally with the AUKUS pact between Australia, the UK, and the U.S. The invitation of India, Indonesia, South Africa, Senegal, and Argentina to 2022's G7 summit reflects another important indicator of a similar trend.

In theory, should the Quad, consisting of Australia, India, Japan, and the United States, merge with AUKUS, it would mark the first alliance of democratic nations ever spanning four continents. Such a specific development is in the long-term not impossible, although it remains a long shot because India most likely has other ideas about its place in the world.

It would be a questionable alliance anyway if centered on military matters. Focusing on commercial and scientific aspects, though, changes near everything. Particularly, if it is about the future of the new Internet: Web3.

Beyond the Military: Web3

Considering how technological platforms underpin most human activities, a predominantly military view has little to offer for the future. Particularly in Asia, it still is too vague and does not materially change the currently existing imbalance that is in China's favor. A more appropriate solution gets much closer to what I am asking for here. It starts with what Graham Allison roughly outlined in a 2020 article about spheres of influence.[275] He mentioned that the U.S. could re-join the Trans-Pacific Partnership (TPP) and extend the agreement to include the European Union.

Should such a system also draw in India, Indonesia, Brazil, and other countries of the group just mentioned, eventually it would include most or all of America's formal and informal allies.

The focus of any such new institution should go beyond mere matters of physical trade. Instead, it should center on the digital world and Web3. Its goal should be to tear down the data-based barriers between businesses and between nations, empower individuals, and enshrine liberal principles in Web3 as the future platform for global commerce and civic activities.

If it could establish and guarantee a genuinely liberal framework for Web3, this would formalize an insurmountable framework linking the activities of billions of people living in a global group of like-minded liberal societies. They would comprise about half of humanity, including some of the most rapidly growing parts of the world. Their economic output would be close to 60% of global GDP.

Web3's adhering to genuinely rules-based and liberal principles, would be the foundation of a 21st century-style liberal order. It could prove the superiority of a decentralized system that trusts its people and inspires their innovative spirits. This would set a direct counterpoint to alternative top-down and authoritarian models such as the one practiced by China.

The before-mentioned hypothetical GAIA group whose establishment Amy Webb outlined to prevent excesses and misuse of digital powers by quasi-monopolies would be an excellent starting point. Its goal is to better align AI and Big Data with individuals' and society's interests. Webb also mentions data security, ownership of personal data, and blockchain governance mechanisms.

Organizations and people living in other nations could also use the system. Even the Chinese could, as is the case today with the Internet. If not, these would not really be effective "commons." But, doing so, all would have to adhere to its liberal rules.

We would have re-created a genuine liberal system for the future. Its digital platform would secure data, protect digital identities and property rights and the free flow of data. This also establishes a genuinely rules-based and liberal platform for commerce. And by sharing data while protecting each person's privacy and security, it would also offer an unsurpassed source for data analytics and AI to help improve our societies.

This would not only set a counterpoint to a "Beijing Consensus" but also to today's often quoted but frequently misleading "rules-based liberal world order." That term too often and too conveniently seems to dismiss the term "liberal" and stress the term "world order." It elevates America's role to the status of a global hegemon. And this militarizes the international role of the United States, unnecessarily so.

China – the Impossible Empire

China is neither Nazi Germany, nor the Soviet Union, and not even the British Empire. It has no ambitions to conquer the world nor to spread its ideology. China wants to be the most powerful country in the world, able to pursue its ambitions without limitations, at least in Asia. At its extreme, this may be world hegemony, but not empire.

But even if world domination would be China's goal, it does not take America's military might to limit and block that path. At least five more critical points must be considered.

First, there are other countries and ambitious nations. In Asia alone, we have Japan, Korea, Indonesia, and the biggest and most noticeably growing challenger: India. Further away, there are the European Union and, of course, North America with the richest large societies in the world.

Second, there are underlying demographic and economic changes. Within the next two to three years already, the population of China will shrink while India and Africa keep expanding, likely even the United States. All these countries' economies are growing, as is Indonesia's.

Third, there is technology, decoupling power ever more from traditional measures of geopolitics. Digitization makes global cooperation and economic growth straightforward and achievable by using brain power instead of physical or kinetic one. Islands and shipping routes are not where the future is won! China can only gain limited geo-strategic benefits even if it were to rule all islands and sea routes from the Paracels to the Senkakus.

Fourth, there are domestic developments in China that may stifle Chinese ambitions. China's meteoric economic rise was above all because of an aggressive laissez-faire and capitalist stance at the heart of its economy. Smart governance, aided by theft and manipulation, contributed as well. But recent developments let us question whether this old recipe has changed and whether the new one is working. Today, a more heavy-handed government limits independent decision-making across society. This reigns in competition and capitalist forces. Demographics, governmental and private debt levels, and massive numbers of retirees with meagre pensions will contribute to shift perceptions, priorities, and likely also outcomes.

Fifth, and above all, there is the liberal world itself, led by the United States. *Over one billion people living in them, close enough to China's numbers, with the gap unlikely to widen substantially*, if at all. Being many times richer than China, this is an economic chasm that is likely to continue for decades. It could even expand. These societies are still leading in most fields of technology and science, the continuation of which they mostly have in their own hands. The richest countries of the liberal world, closely aligned and almost all allied, can hold their own.

Winning Without War

When China (also) realizes that the very nature of geopolitics has changed and is being transformed, and that new and decisively democratic and effective liberal rules apply around the globe and beyond, baked into the core platforms of the future, the calculus of its leaders will change. They may still consider striking out at Taiwan. But then, they also must be much more concerned about losing out in an area that matters much more to the Chinese population: the economy. And just maybe, they then revert to doing what successful Chinese governments have done in the past: deliver prosperity to their people. Hardly anything can be more important to Chinese leaders, because history shows that lacking to do so will lead to discontent, sometimes even revolution and the deposition of failed rulers.

China's reaction and strategies may be called Communism with Chinese characteristics, or Capitalism with Communist characteristics, or Chinese Communism with Capitalist characteristics - or simply a re-emerged China. It would all be acceptable enough with me, at least compared to a world war.

As I laid out above, not just us but also China is going to live in a multi-polar world, even on the Eurasian landmass. Unless some secret sauce and unique leverage of technology will dramatically alter the power balance (Quantum and AI, maybe?), China will have an unsurmountable hurdle to project enough power into the rest of Earth to dominate the world.

And it is implausible to assume that China could change this calculus. There always will be India, the U.S., and Europe, and arguably also Russia and other emerging nations in Africa, Asia, and Latin America that have not yet reached their full potential either.

The world will coalesce around our liberal system, and actively support it, as long as we ensure we can protect the new spatial and digital commons while delivering superior solutions to our people. We can enable the future and save our world - without destroying it.

The United States of America has been unique among all major societies in the world in that it never has been ravaged by foreign wars, defeat, or inflation.

Let's keep it that way.

EPILOGUE: AMERICA'S CHOICE

And Now For Something Completely Different...

If we make it through the 2020s without a bloody war between China and the U.S. that spills into our homelands, the dynamics are likely to change. By then, China's population will shrink and age faster than the one of a growing America. China's economy will probably have passed the United States by any measure. But India will have started to emerge as a new disruptor on the world stage. Economic and technological developments will have given us much more clarity over the shape of Web3 and the new world's digital metaverses. And we will also better understand which society has managed to best take advantage of tech-induced opportunities for wealth creation. We will know who and how they addressed and solved urgent issues of health, education, pensions, environment, and individual opportunity.

First, though, we must get there. For this, we need restraint, humility, and critically: diplomacy and strategic vision.

In the before-mentioned 2020 article in Foreign Policy, Harvard Professor Graham Allison advocates for re-introducing the concept of spheres of influence into international relations. He argues we must ask ourselves what our core interests are and then decisively act on our answer. The rationale laid out in his article suggests that he sees Taiwan as within the PRC's sphere of influence and that we should not fight over it.

Former political affairs officer George Paik's answer is different, pointing out that a U.S. that takes its founding principles seriously must be prepared to militarily defend the genuinely free society of Taiwan.[276] Both authors, though, do not clearly express the exact thresholds. As so often, the devil is in the details. In the first case, would the U.S. abrogate any support for Taiwan? And in the second, would America defend Taiwan with all its military might and to the same degree as if it were a treaty ally, or even part of its own territory?

Hugh White, Emeritus Professor of Strategic Studies at the Australian National University, contemplates such questions, although mostly from the perspective of U.S. ally Australia.[277] Like Allison, he clearly sees actively fighting for Taiwan as, at best, a noble gesture, but in reality a losing proposition that we should not seriously consider. He seems aligned in this with Australia's former premier minister, Kevin Rudd.

What, though, can we do to prevent conflict in the first place?

South China Sea and Small Islands

And this then brings us back to spheres of influence and diplomacy. Above all, this warrants another look at the South China Sea (SCS). There, China has resisted a multinational solution but insisted on bilateral negotiations or even more forcefully the unilateral establishment of facts, of a fait accompli. The latter is why it has militarily occupied several islands, only verbally opposed by the Obama Administration.

To cut this Gordian Knot, we must geopolitically decouple China's First Island Chain (roughly all islands South of Okinawa that lie within 200 miles of the Chinese mainland) from the West's First Island Chain and from the Southern half of the SCS. Inside China's First Island Chain, almost all trade is Chinese trade. It is very close to China. The islands are mostly unpopulated. Alternative historical claims do not obviously seem better than China's.

We must consider one of the core points of this book, the shift from geography to the digital worlds. If we then exclude Taiwan, from the Western allies' perspective, there are mostly pride and symbolic values involved. At some point, the Alsace also seemed worth fighting major wars over, just to discover a mere generation later that it could *unite* two former enemies rather than divide them.

We must continue to resist the PRC's claim of ownership over the Spratly Islands, though. These are too far south and too close to the waters of other claimants. They also directly and significantly limit shipping lanes and turn century-old conventions on their head based on vague claims of much smaller-scale use by some fishing boats of ancient empires.

Likewise, the Scarborough Shoals and many other reefs are too close to the Philippines, or in the very south also to Malaysia, Brunei, and Vietnam. A much better way of handling these features is to either subject them to international rulings or to agree among all claimants. All these islands should be de-militarized. Once the ground rules are agreed upon, their use for fishing and mining, and their ecological protection, could be put under a common supranational agency. This should assure freedom of navigation.

The Paracels Islands are different. They are so close to the Chinese mainland that putting them under the PRC's administrative control does not sound unreasonable. The specifics, together with those for the rest of the SCS, would need to be defined in a joint agreement. A bigger question is whether they should be de-militarized. We should aim for that, although this depends on the overall agreement.

The result would be a demilitarization of most of the SCS, formal Chinese control and exploration rights over the Paracels, with joint management of the other islands. Such management may be through a separate agency, through a body administered by the UN, or through ASEAN.

This means a deal. And why not? It simply would enable all participants to compromise and get most of what they think they *need*, while all would also lose something they *want*. Everybody would get clarity and have their most urgent territorial claims or worries satisfactorily addressed. It would eliminate one of the biggest military hotspots on planet Earth.

The main and obvious question is why China would agree to such an arrangement.

When one looks at the overall picture, it would enable China to establish an internationally accepted Air Declaration Zone (ADZ) over the Paracels and prevent FONOPs in this most sensitive area close to the Chinese coast and to the PLAN base housing its Southern Command. This would not involve any significant concessions on either side. A framework and agency addressing ecological and disaster management would further reduce the risk of conflict. Chinese commercial activities would, in principle, not be limited. But its activities would no longer be seen as threatening. The outlined SCS solution would even openly show that the PRC will work constructively with regional multilateral bodies like ASEAN.[278]

And besides, it could be part of a larger deal, one that may not quite be about "spheres of influence" but establish clarity across all of East and Southeast Asia. This book is not the place to get into details about the Senkakus, which are disputed by Japan, Taiwan, and the PRC. The U.S. states that it would defend the islands because they are currently administered by Japan, while in principle being neutral on the question of who has the actual sovereignty over these islands.

But we must, once more, talk about Taiwan.

A Role for Diplomacy?

Unfortunately, with nearly anything along China's coast, military aspects dominate the conversation and the actions on both sides. Old-school diplomacy, it seems, has tragically moved far out of the picture. As a result, we lack genuinely creative initiatives that go beyond pride and ideological principles.

Why can we not find a peaceful solution for the Taiwan question? Today already it has a special status. This is an indisputable fact. Can't we explore a middle ground, interpreting what "two systems" means? And then put it into action? Is it absolutely unrealistic? Taiwan could be an outpost and entry point into China for us, and a link for the PRC to the West. There are models in history. Free Trade Zones, united kingdoms, autonomous regions, and more. The above-mentioned digital-focused new liberal system and platforms could be a means in that process.

Of course, we have seen what China did in Hong Kong, a mere 25 years after the formal agreement to transfer control. But Taiwan is not Hong Kong, or at least it does not have to be. And Hong Kong had a quarter century to adjust. A comparable time span would get us close to the middle of the 21st century. The world will look profoundly different by then, for all parties involved.

What China Wants from Us...

Diplomacy has a chance because *there is a lot China wants from us*. And not just the PRC, but also President Xi wants that. For example, it wants achievement of its most important naval aspirations, with full control over disputed islands and foreign navies out of waters close to its coasts. The PRC also wants elimination of conflict zones, friendly relationships with neighbors, geopolitical security, uninhibited access to the blue oceans, access to markets and global commerce, clarification of the political status of Taiwan, Tibet, and Xinjiang, prevention of war, stop of weapons deliveries to Taiwan, and even just a reduced risk to lose much in a war over Taiwan. Agreeing to many of these points, probably most, would not change any de facto status. None of this is outside the greater area of influence that China already has established, and none of this would result in any material concession by America or its treaty allies.

For example, what about granting China a sphere of influence over - China? We can disagree about each other's domestic actions, and aggressively denounce brutal re-education camps, cultural cleansing, and forced sterilizations in Xinjiang without implicitly questioning China's political control over its western regions, including Tibet. And if the Chinese really are confident and believe in their actions, it may bother them, but they must accept us talking about it.

And China can continue criticizing inefficient and inconsistent Western practices, domestically and internationally.

Either way, we can decouple this.

... and What We Want from China

There also is a lot that we want from China. We want access to its markets, stop interfering with our domestic affairs, letting our businesses and their employees speak freely, stop intellectual espionage and hacking, an open bi-directional flow of technological and scientific advances, free commerce and open competition, settled borders for China's neighbors, prevention of conflict with us and our closest friends, and the elimination of meddling in international organizations.

Or what about genuinely global issues, relating to the environment, climate, infectious diseases, space, radio frequencies, or security of technology?

Have we seriously enough explored arms control yet? Everybody keeps counting numbers of missiles and planes, and tonnage of blue ocean navies, pointing out their price. Is our only option trying to out-build the other side? We talk about abilities to intercept satellites, but what about hardening each other's systems? Or turning satellite systems, and digital pathways into the open seas of the near future?

There are also serious things both sides could lose. Millions of human lives, for example, as outlined in Chapter 12 above. Or our prosperity.

Taiwan can divide us or bring us together. I am not comparing Taiwan to a hill in the Southern Alsace, although war over it would in some ways be similar. Taiwan can be like the Alsace is today, a connecting point and bridge, with its own cultural context.

And a Role for ASEAN

Specifically in East and Southeast Asia, we also must directly address the current areas of potential conflict, from Japan to Singapore. Without reducing military cooperation with friends like the Philippines, Singapore, Thailand, Malaysia, or Indonesia, the U.S. should also more actively engage with ASEAN. The organization can be a bridge, commercially, and for ideas and political engagement. At least since President Obama's "pivot" to Asia we have talked about stuff like that. What about talking less and acting more? And not using the military to do either?

By granting a large role to a regional supra-national entity with ten diverse members, this removes most of the matters addressed by this organization from direct geopolitical competition. It reduces China's disproportional leverage as well.

We also must lean much stronger on the dominant power for all trade between the Pacific and the Indian Ocean. Indonesia has almost 300 million people living in a booming economy. This archipelago nation is the default gateway and guardian of all routes connecting the Pacific and the Indian Ocean. It would make an outstanding addition to any framework that protects liberal values.

Indonesia has another added benefit. It refuses to being dominated by either China or India. And this provides a balance that soon may become more important. China exerts itself today, but India also has its own aspirations, and in the medium turn it will turn into a power with its own more assertive global aspirations.

The result of the just outlined diplomatic solutions is that it would liberate Chinese physical trade and activities in line with the PRC's long-standing aspirations. Its trade into the Indian Ocean and from there into the Atlantic, though, would be reasonably balanced by gatekeepers that are neutral (ASEAN) or friendly to the West (Japan, Philippines).

Today, though, we have not even started to give diplomacy a real chance.

Please review this book.

And please give feedback via:

https://WarOrWealth.com

Acronyms

A2/AD	Anti-Access/Area Denial
ADZ	Air-Declaration Zone
AFRL	Air Force Research Laboratory
AG	Aktiengesellschaft
AI	Artificial Intelligence
AIP	American Institute of Physics
APS	American Physics Society
AR	Artificial Intelligence
ASEAN	Association of South East Asian Nations
AUKUS	Australia, U.K., U.S.
BJP	Bharativa Janata Party
BLM	Black Lives Matter
BP	British Petroleum
BRI	Belt and Road Initiative
BRICS	Brazil, Russia, India, China, South Africa
C3	Communication, Command, and Control
CCP	Chinese Communist Party
CDC	Center for Disease Control
CEO	Chief Executive Officer
CIA	Central Intelligence Agency, also: Confidentiality, Integrity, Availability
CNN	Cable News Network
CO2	Carbon Dioxide

COVID	Corona-Virus Disease
CSG	Carrier Strike Group
DARPA	Defense Advanced Research Projects Agency
DC	District of Columbia
DDOS	Distributed Denial of Service
DF	Dong Feng
DIU	Defense Intelligence Unit
DLT	Digital Ledger Technlogies
DNA	Deoxyribonucleic Acid
DOD	Department of Defense
DSRC	Dedicated Short-Range Communications
EMP	Electro-Magnetic Pulse
EU	European Union
ESG	Environmental, Social, Governance
FONOP	Freedom-of-Navigation Operations
GAIA	Global Alliance for Intelligence Augmentation
GDP	Gross Domestic Product
GEO	Geostationary Earth Orbit
GMO	Genetically Modified Organism
GPS	Global Positioning System
GRE	Graduate Record Examination
IAPP	International Association of Privacy Professionals
ICE	Immigration and Customs Enforcement
IED	Improvised Explosive Device
INF	Intermediate-Range Nuclear Forces

IoT	Internet of Things
IIoT	Industrial Internet of Things
IP	Intellectual Property
IPCC	Intergovernmental Panel on Climate Change
ISIS	Islamic State of Iraq and Syria
ISP	Internet Service Provider
JSF	Joint Strike Fighter
LEO	Low-Earth Orbit
MIT	Massachusetts Institute of Technology
NATO	North-Atlantic Treaty Organization
NSA	National Security Agency
NSF	National Science Foundation
OECD	Organization for Economic Co-operation and Development
OPM	Office of Personnel Management
P2P	Peer-to-Peer [Communications]
PLA	People's Liberation Army
PLAN	People's Liberation Army's Navy
PLARF	People's Liberation Army Rocket Force
PPP	Purchase Power Parity
PRC	People's Republic of China
QKD	Quantum Key Distribution
R&D	Research and Development
RSIS	Rajaratnam School of International Studies
S&E	Science and Engineering
SAR	Synthetic-Aperture Radar
SBSP	Space-Based Solar Power

SCS	South-China Sea
SMB	Small and Medium-sized Businesses
STEM	Science, Technology, Engineering, Math
TPP	Trans-Pacific Partnership
TV	Television
UAE	United Arab Emirates
UN	United Nations
UK	United Kingdom
UPS	United Parcel Service
USPS	United States Postal Service
U.S.	United States (of America)
V2G	Vehicle-to-Grid
V2V	Vehicle-to-Vehicle
VC	Venture Capital
VLEO	Very Low-Earth Orbit
WHO	World Health Organization
WiFi	Wireless Fidelity

Bibliography

The following are selective books addressing most of the topics covered in this book.

Author(s)	Published	Title
Allison, Graham	2017	Destined for War: Can America and China Escape Thucydides' Trap?
Bricker, Darrell; Ibbitson, John	2019	Empty Planet: The Shock of Global Population Decline
Brose, Christian	2020	The Kill Chain: Defending America in the Future of High-Tech Warfare
Cain, Geoffrey	2021	The Perfect Police State: An Undercover Odyssey into China's Terrifying Surveillance Dystopia of the Future
Clarke, Richard A; Knake, Robert K.	2019	The Fifth Domain: Defending Our Country, Our Country, and Ourselves
Colby, Elbridge A.	2021	The Strategy of Denial
Collins, John M.	1998	Military Geography for Professionals and the Public
De la Cruz, Diego Sánchez	2017	Por qué soy liberal
Diamandis, Peter; Kotler, Steven	2020	The Future is Faster Than You Think

Author(s)	Published	Title
Etzioni, Amitai	2017	Avoiding War with China: Two Nations, One World
Gilder, George	2016	The Scandal of Money
Gilder, George	2018	Life After Google
Haskel, Jonathan; Westlake, Stian	2017	Capitalism Without Capital: The Rise of the Intangible Economy
Kissinger, Henry	2011	On China
Kissinger, Henry A.; Schmidt, Eric; Huttenlocher, Daniel	2021	The Age of AI: And Our Human Future
Lee, Kai-Fu	2018	AI Superpowers: China, Silicon Valley, and the New World Order
Leebaert, Derek	2010	Magic and Mayhem: The Delusions of American Foreign Policy From Korea to Afghanistan
Mearsheimer, John J.	2003	The Tragedy of Great Power Politics
Morland, Paul	2019	The Human Tide: How Population Shaped the Modern World
Nye, Joseph S.	2015	Is the American Century Over?
Nye, Joseph S. Nye	2011	The Future of Power
O'Hanlon, Michael	2019	The Senkaku Paradox: Risking Great Power War Over Small Stakes

Author(s)	Published	Title
O'Hanlon, Michael	2021	The Art of War in an Age of Peace: U.S. Grand Strategy and Resolute Restraint
Preble, Christopher A.	2019	Peace, War, and Liberty
Pry, Dr. Peter Vincent	2021	Blackout Warfare: Attacking the U.S. Power Grid - A Revolution in Military Affairs
René, Gabriel; Mapes, Dan	2019	The Spatial Web: How Web 3.0 Will Connect Humans, Machines, and AI to Transform the World
Rosling, Hans; Rosling; Anna Rönnlund; Rosling, Ola	2018	Factfulness: Ten Reasons We Are Wrong About the World - and Why Things Are Better Than You Think
Rubin, Peter	2018	Future Presence: How Virtual Reality Is Changing Human Connection, Intimacy, and the Limits of Ordinary Life
Rudd, Kevin	2022	The Avoidable War: The Dangers of a Catastrophic Conflict Between the US and Xi Jinping's China
Scharre, Paul	2018	Army of None: Autonomous Weapons and the Future of War
Schwab, Klaus; Davis, Nicholas	2018	Shaping the Future of the Fourth Industrial Revolution
Shambaugh, David	2016	China's Future
Smil, Vaclav	2018	Energy and Civilization: A History

Author(s)	Published	Title
Smil, Vaclav	2021	Grand Transitions: How the Modern World Was Made
Smil, Vaclav	2020	Growth: From Microorganisms to Megacities
Surowiecki, James	2005	The Wisdom of Crowds
Szu-Chien, Hsu; Cole, Michael	2020	Insidious Power: How China Undermines Global Democracy
Taylor, Brendan	2018	The Four Flashpoints: How Asia Goes to War
Webb, Amy	2019	The Big Nine: How the Tech Titans and Their Thinking Machines Could Warp Humanity
Weichert, Brandon J.	2020	Winning Space: How America Remains a Superpower
White, Hugh	2013	The China Choice: Why We Should Share Power
White, Hugh	2019	How to Defend Australia
Wittes, Benjamin; Blum, Gabriella	2015	The Future of Violence: Robots and Germs, Hackers and Drones - Confronting A New Age of Threat
Zeihan, Peter	2020	Disunited Nations: The Scramble for Power in an Ungoverned World
Zeihan, Peter	2022	The End of the World is Just the Beginning: Mapping the Collapse of Globalization

Author(s)	Published	Title
Zuboff, Shoshana	2019	The Age of Surveillance Capitalism: The Fight for a Human Future at the New Frontier of Power

Notes

The below notes concentrate on providing explanations and references to the most relevant facts mentioned in this book. I made a judgement call of not supporting every single statement or number, particularly not those that are not contentious and can be assumed generally recognized public knowledge. The dates in brackets refer to the last date that web links were verified to be valid.

---

1 https://www.fpri.org/article/2017/10/parsing-xi-jinpings-statements-taiwan-19th-communist-party-congress/ [9 Jul 2022]

2 https://www.visualcapitalist.com/2000-years-economic-history-one-chart/ [5 Aug 2022]

3 The jump between GPT-3 and GPT-4 is one of these examples. See: https://www.forbes.com/sites/bernardmarr/2021/09/24/the-7-biggest-artificial-intelligence-ai-trends-in-2022/?sh=5ba012662015

4 More about their points later.

5 https://www.memorial-hwk.eu/en/battlefield

6 Kissinger, Henry A.; Schmidt, Eric; Huttenlocher, Daniel (2021), The Age of AI And our Human Future; https://www.amazon.com/Age-I-Our-Human-Future/dp/0316273805

7 https://dod.defense.gov/Portals/1/Documents/pubs/ASB-ConceptImplementation-Summary-May-2013.pdf [3 Apr 2022]

8 In great detail laid out by Harvard's Shoshana Zuboff in her 2019 "The Age of Surveillance Capitalism." More about that later.

[9] According to CDC statistics, both of these numbers are 30 times as high, with over 30,000 each.
https://www.cdc.gov/nchs/fastats/accidental-injury.htm [3 Apr 2022]

[10] In Chapter 8 of the following book, Stephanie Kelton describes how this works today without much partisan concern, unlike other priorities of society. Of course there is one consequence of simply creating more money: inflation.
Kelton, Stephanie, The Deficit Myth: Modern Monetary Theory and the Birth of the People's Economy (2020, Hachette Book Group)
https://www.amazon.com/Deficit-Myth-Monetary-Peoples-Economy-ebook/dp/B07RM72BT7/

[11] In his 2001 book, "Does America Need a Foreign Policy," Henry Kissinger goes into great detail in the roots and challenges with America's normative foreign policy.
Kissinger, Henry (2001), Does America Need a Foreign Policy? : Toward a New Diplomacy for the 21st Century;
https://www.amazon.com/Does-America-Need-Foreign-Policy-ebook/dp/B000FC0NKA/ref=tmm_kin_swatch_0?_encoding=UTF8&qid=1649351946&sr=8-1 [7 Apr 2022]

[12] The last indisputable great power war was World War II, over 75 years ago. China and the U.S. fought each other during the Korean War between 1950 and 1951, with some subsequent battles until the 1953 armistice. That war, though, remained limited in its geographical scope. Also, it can be questioned whether China was a genuine "great power." At least theoretically, the U.S. had such a significant technological and economic advantage that it could have destroyed the People's Republic of China without China being able to retaliate to a comparable degree.

[13] The groundwork for some of the tactics and strategies of World War I were laid way after the Napoleonic wars, i.e. in the Crimean, Austrian-Prussian, and even far away in the U.S. Civil War.

[14] For example, the budget reflects tens of billions of dollars for research, development, and testing, but no significant amounts to put many of these systems into practice for real-life feedback. https://www.csis.org/analysis/analysis-fy-2022-defense-budget [3 Apr 2022]

[15] Winston Churchill said on Nov 11, 1947, "Many forms of Government have been tried, and will be tried in this world of sin and woe. No one pretends that democracy is perfect or all-wise. Indeed it has been said that democracy is the worst form of Government except for all those other forms that have been tried from time to time...." https://winstonchurchill.org/resources/quotes/the-worst-form-of-government/ [11 Jul 2022]

[16] https://www.wsj.com/articles/subsidies-chips-china-state-aid-biden-11627565906 [12 Jul 2022]

[17] https://www.forbes.com/sites/rainerzitelmann/2019/07/08/chinas-economic-success-proves-the-power-of-capitalism/?sh=7eae82f43b9d [13 Jul 2022]

[18] These companies' founders manage to keep effective control by introducing different classes of shareholder and keeping majority stakes of *voting* shares for themselves.

[19] Again, one of the most detailed accounts putting this into context is in Shoshana Zuboff's "The Age of Surveillance Capitalism."

[20] 2022 numbers published by the U.S. census bureau show 1,410,539,758 for China and 23,580,712 for Taiwan. This translates to Taiwan's population being 1.67% of the number of China. https://www.census.gov/popclock/world [8 Jul 2022]

---

[21] One typical example is the following paper written by Eric Maskin at Harvard University, about such claims by Hayek, as compared to the deliberate construction of more effective economic institutions via "mechanism design."
https://scholar.harvard.edu/files/maskin/files/friedrich_von_hayek_and_mechanism_design_e._maskin.pdf [8 Jul 2022]

[22] The World Bank shows $15.05 trillion for the U.S. in 2010 and $21.37 trillion in 2020, with the equivalent numbers for China being $6.09 trillion and $14.28 trillion.
https://data.worldbank.org/indicator/NY.GDP.MKTP.CD?locations=CN-US-JP [8 Jul 2022]

[23] The totals are slightly below 58% for East and Southeast Asia, and above 46% for all of Asia, in both cases excluding Russia.
https://data.worldbank.org/indicator/NY.GDP.MKTP.CD?locations=CN-8S-Z4-ZQ-Z7 [30 Mar 2022]

[24] https://www.britannica.com/topic/geopolitics [3 Apr 2022]

[25] https://www.merriam-webster.com/dictionary/geopolitics [3 Apr 2022]

[26] Rudd, Kevin (2022), The Avoidable War: The Dangers of a Catastrophic Conflict Between the US and Xi Jinping's China; https://www.amazon.com/Avoidable-War-Dangers-Catastrophic-Conflict-ebook/dp/B08ZMSM93S/

[27] https://www.military.com/daily-news/2020/09/01/chinas-military-has-surpassed-us-ships-missiles-and-air-defense-dod-report-finds.html [7 Apr 2022]

[28] As of 2020, statistics show 42.24 % as Han ethnicity, 44.96 % as Uygur ethnicity, and 12.8 % as other ethnic minorities; https://www.globaltimes.cn/page/202106/1226080.shtml [8 Jul 2022]

[29] In Tibet, this process is not as advanced yet. It started on a much lower level, indicating how different the actual cultures really were. https://www.niussp.org/migration-and-foreigners/han-chinese-population-shares-in-tibet-early-insights-from-the-2020-census-of-china/ [8 Jul 2022]

[30] https://www.reuters.com/article/india-iran-ports-int-idUSKBN2AX1DK [8 Jul 2022]

[31] https://www.cfr.org/blog/belt-and-road-initiative-forcing-europe-reckon-china [8 Jul 2022]

[32] "Prisoners of Geography" by Tim Marshall, Chapter 2. See: https://www.amazon.com/Prisoners-Geography-Explain-Everything-Politics/dp/1501121472 [3 Apr 2022]

[33] The World Economic Forum states the share of sea-borne trade to be about 90% of the world's total. https://www.weforum.org/agenda/2021/10/global-shortagof-shipping-containers/ [8 Jul 2022]

[34] This can be inferred out of the following UNCTAD report. https://unctad.org/system/files/official-document/rmt2021_en_0.pdf [8 Jul 2022]

[35] According to an arguably more relevant definition, purchasing power parity or PPP, the World Bank even considers it the world's biggest.

[36] Rudd, Kevin (2022), The Avoidable War: The Dangers of a Catastrophic Conflict Between the US and Xi Jinping's China; Chapter 8

[37] https://www.forbes.com/sites/davidaxe/2021/11/05/yes-china-has-more-warships-than-the-usa-thats-because-chinese-ships-are-small/?sh=5f43bccb611d [3 Apr 2022]

[38] https://indianexpress.com/article/india/60-per-cent-navy-forces-in-indo-pacific-region-now-us-navy-chief-7568984/ [8 Jul 2022]

---

[39] According to the South China Morning Post. https://www.businessinsider.com/chinese-military-could-be-ready-to-challenge-us-in-indo-pacific-by-2035-report-2018-11?op=1 [8 Jul 2022]

[40] For various technologies, see the following articles. https://www.sciencedirect.com/science/article/abs/pii/S0141118718300014 https://www.nti.org/analysis/articles/submarine-detection-and-monitoring-open-source-tools-and-technologies/ https://www.scmp.com/news/china/science/article/3025096/chinese-us-scientists-develop-ai-technology-help-detect [8 Jul 2022]

[41] The operating depth of many U.S. submarines is around 3,000 feet. https://navalpost.com/how-deep-can-a-submarine-dive/ [8 Jul 2022]

[42] The rate of digitization is more than doubling within five years to over $1 trillion by 2025, and the capabilities of local micro factories are growing rapidly. https://www.globenewswire.com/news-release/2020/08/14/2078517/0/en/The-World-s-Digital-Transformation-Industry-2020-2025-Trends-Opportunities-and-Competitive-Landscape.html and https://www.weforum.org/agenda/2019/06/localized-micro-factories-entrepreneurs-and-consumers/ [8 Jul 2022]

[43] Many of the following data are based on two recent books by experts on demographics, and their verified links to publicly accessible scientific studies: Morland, Paul (2019), The Human Tide: How Populations Shaped the Modern World; https://www.amazon.com/Human-Tide-Population-Shaped-Modern-ebook/dp/B07F6B3D5F/ [7 Apr 2022]
Bricker, Darrell; Ibbitson, John (2019), Empty Planet: The Shock of Global Population Decline; https://www.amazon.com/Empty-Planet-Global-Population-Decline-ebook/dp/B07CWHYVW5/ [7 Apr 2022]

[44] Other sources put the number lower, suggesting either 2024 or 2023 as the year. https://www.newsweek.com/when-will-india-overtake-china-population-1590451 [28 Jul 2022]

[45] Empty Planet, Chapter 4, page 153: Wolfgang Lutz' predictions about China are based on his Rapid Development Model. Additional data is from the Chinese State Statistical Bureau 2016

[46] https://www.scientificamerican.com/article/there-are-more-boys-than-girls/ [28 Jul 2022]

[47] Obviously, the latter can and often do impact the former. Governmental policies furthering education and ensuring access to health care, changing cultural values, and a sustainable economic situation contribute to women's education.

[48] According to the U.S. Census Bureau. https://www.census.gov/content/dam/Census/library/working-papers/2009/demo/us-pop-proj-2000-2050/analytical-document09.pdf

[49] https://teacher.scholastic.com/activities/immigration/immigration_data/index.html [7 Apr 2022]

[50] This calculation of 19th century immigration are based on census results and the just mentioned immigration numbers per decade. See also: https://www.thoughtco.com/us-population-through-history-1435268 [7 Apr 2022]

[51] Yglesias, Matthew (2020), One Billion Americans: The Case for Thinking Bigger; https://www.amazon.com/One-Billion-Americans-Thinking-Bigger-ebook/dp/B082ZR6827/ [7 Apr 2022]

[52] One of many examples is this article in which he makes his point concisely. https://mahbubani.net/the-u-s-must-approach-china-with-humility-newsweek-opinion/ [13 Jul 2022]

[53] Indirectly related, an interesting although somewhat dense book introduces the concept of quantum theory into the world of social

sciences. See: Wendt, Alexander, Quantum Mind and Social Science: Unifying Physical and Social Ontology (2015) - https://www.amazon.com/Quantum-Mind-Social-Science-Unifying-ebook-dp-B00U7CRGEE/dp/B00U7CRGEE/. [13 Jul 2022]

[54] Department of Economic and Social Affairs, World Population Prospects: 2012 Revision, June 2013, http://esa.un.org/unpd/wpp/index.htm [7 Apr 2022]

[55] China is getting close to the U.S. in R&D spending, and a comparable shortage of STEM workers (less than half of the Chinese number of bachelor degrees in STEM). https://ncses.nsf.gov/pubs/nsb20201/global-r-d and https://www.aip.org/fyi/2019/panel-warns-us-faces-stem-workforce-supply-challenges and https://www.aip.org/fyi/2018/rapid-rise-chinas-stem-workforce-charted-national-science-board-report [8 Jul 2022]

[56] In 2020, the gross domestic savings in China were 44.7%, compared to 18% in the U.S. https://data.worldbank.org/indicator/NY.GDS.TOTL.ZS?locations=CN-US [8 Jul 2022]

[57] Sinclair, David (2019), Lifespan: Why We Age - And Why We Don't Have To; https://www.amazon.com/Lifespan-Why-Age-Dont-Have-ebook/dp/B07N4C6LGR/ [7 Apr 2022]
And Robbins, Tony; Diamandis, Peter (2022), Life Force: How New Breakthroughs in Precision Medicine Can Transform the Quality of Your Life & Those You Love; https://www.amazon.com/Life-Force-Breakthroughs-Precision-Transform-ebook/dp/B09KY7ZXV6/ [7 Apr 2022]

[58] Deutsch, Breanna (2020), Finding the Fountain: Why Government Must Unlock Biotech's Potential to Maximize Longevity;

https://www.amazon.com/Finding-Fountain-Government-Potential-Longevity-ebook/dp/B08L8T2N8Y/ [7 Apr 2022]

[59] Among them are the above-mentioned David Sinclair from Harvard University, George Church at Harvard University, Kris Verburgh of the University of Brussels, and Andrew Steele of the Crick Institute in London.

[60] https://www.buckinstitute.org

[61] https://colossal.com

[62] https://population.un.org/wpp/Graphs/Probabilistic/POP/TOT/643 [7 Apr 2022]

[63] More specifically, although still not 100% exact, is to use the number 72 instead of 70. Therefore, this rule is also known as the "rule of 72."

[64] More specifically, his 1990 "24-Character Strategy" advised to, "Observe calmly; secure our position; cope with affairs calmly; hide our capacities and bide our time; be good at maintaining a low profile; and never claim leadership." https://www.globalsecurity.org/military/world/china/24-character.htm [3 Apr 2022]

[65] https://seekingalpha.com/article/220963-china-surpassed-japan-as-2-economic-nation-apparently [3 Apr 2022]

[66] It is almost three times as big, in nominal GDP. https://databank.worldbank.org/data/download/GDP.pdf [3 Apr 2022]

[67] The OECD and the Japan Center for Economic Research put the number a bit higher, at slightly above 10%, but the trend line remains unchanged:
https://www.oecd.org/economy/outlook/2060policynote.pdf
https://www.jcer.or.jp/english/2060-digital-global-economy [8 Jul 2022]

---

[68] By Marshall Goldsmith and Mark Reiter. See: https://www.amazon.com/What-Got-Here-Wont-There/dp/0739342231 [3 Apr 2022]

[69] And purely economically speaking the European Union. The EU has just not yet sufficiently proven its cohesion when the U.S. military shield is taken away.

[70] https://statisticstimes.com/economy/united-states-vs-china-economy.php [7 Apr 2022]

[71] https://statisticstimes.com/economy/united-states-vs-china-economy.php [7 Apr 2022]

[72] These happened during times of wild swings and during relative short time spans, but it is still notable. https://www.wsj.com/articles/u-s-economy-likely-to-outgrow-chinas-due-to-contrast-in-pandemic-responses-11629036000 [8 Jul 2022]

[73] Peter Zeihan wrote a bestseller that goes into great detail on this topic, although it also seems somewhat simplistic and sensationalist. See: Zeihan, Peter, The End of the World is Just the Beginning: Mapping the Collapse of Globalization (2022); https://www.amazon.com/End-World-Just-Beginning-Globalization-ebook/dp/B09C65JNPF/

[74] According to a march 2015 Policy Brief of the Peterson Institute for International Economics, by Jakob Funk Kierkegaard, named "The True Levels of Government and Social Expenditures in Advanced Economies https://www.piie.com/publications/pb/pb15-4.pdf [8 Jul 2022]

[75] https://theconversation.com/to-get-rich-is-glorious-how-deng-xiaoping-set-china-on-a-path-to-rule-the-world-156836 [22 Jun 2022]

[76] This has preceded the Covid-19 pandemic. https://www.cnbc.com/2019/07/09/us-life-expectancy-has-been-declining-heres-why.html [8 Jul 2022]

[77] https://www.newsweek.com/how-chinese-communist-party-outwitted-american-capitalists-1605835 [23 Jun 2022]

[78] https://www.cnbc.com/2017/03/02/chinas-parliament-has-about-100-billionaires-according-to-data-from-the-hurun-report.html [23 Jun 2022]

[79] See: https://www.nytimes.com/2016/09/04/business/international/venture-communism-how-china-is-building-a-start-up-boom.html and more recently also https://www.bloomberg.com/news/features/2022-02-06/where-is-china-investing-communist-leaders-are-becoming-venture-capitalists

[80] Lee, Kai-Fu (2018), AI Superpowers: China, Silicon Valley, and the New World Order; https://www.amazon.com/AI-Superpowers-China-Silicon-Valley-ebook/dp/B0795DNWCF/ [7 Apr 2022]

[81] https://wits.worldbank.org/CountryProfile/en/Country/CHN/Year/LTST/TradeFlow/Export/Partner/all/ [7 Apr 2022]

[82] This seems to be a reference to the 170 member "Standing Committee of the National People's Congress." https://www.nytimes.com/2018/03/01/business/china-parliament-billionaires.html [9 Jul 2022]

[83] https://www.forbes.com/sites/giacomotognini/2021/04/06/the-countries-with-the-most-billionaires-2021/?sh=2a0109a4379b [9 Jul 2022]

[84] See for example the 2014 forecast based on 2010 PPP numbers published by the OECD. https://data.oecd.org/gdp/gdp-long-term-forecast.htm and PwC's 2017 report "The Long View" at https://www.pwc.com/gx/en/research-insights/economy/the-world-in-2050.html [8 Jul 2022]

---

[85] https://www.jcer.or.jp/english/2060-digital-global-economy [13 Jul 2022].

[86] https://www.jcer.or.jp/jcer_download_log.php?f=eyJwb3N0X2lkIjo1ODI5OCwiZmlsZV9wb3N0X2lkIjo1ODMxOX0=&post_id=58298&file_post_id=58319 [8 Jul 2022]

[87] https://www.newsweek.com/us-professor-fired-chinese-university-1030284 [13 Jul 2013]

[88] https://www.brookings.edu/bpea-articles/a-forensic-examination-of-chinas-national-accounts/ [13 Jul 2022] and https://nationalinterest.org/blog/buzz/surprise-chinas-economy-smaller-you-think-127042 [13 Jul 2022]

[89] https://www.nytimes.com/2020/12/24/technology/china-jack-ma-alibaba.html [8 Jul 2022]

[90] I must point out that, as I will describe later in the book, that I do acknowledge shortcomings of markets. These are similar to those measuring "GDP" as a means of economic capability. In my opinion, some markets can be optimized by splitting them and linking them to decentralized mechanisms similar to blockchains that focus on maximizing socially acceptable outcomes. This is particularly the case for markets that are, often for a good political reasons, highly regulated, like energy, education, and healthcare.

[91] Nye, Joseph (2005), Soft Power: The Means to Success in the World; https://www.amazon.com/Soft-Power-Means-Success-Politics-ebook/dp/B005GKIY2U/ [7 Apr 2022]

[92] Helper, Stefan (2012), The Beijing Consensus: Legitimizing Authoritarianism in Our Time; https://www.amazon.com/Beijing-Consensus-Legitimizing-Authoritarianism-Time-ebook/dp/B0073SYA1C/ [7 Apr 2022)]

[93] Indonesia later started to shift its position. https://thediplomat.com/2022/04/which-asian-countries-voted-to-suspend-russias-unhrc-membership/ [10 Apr 2022]

[94] https://www.theglobeandmail.com/world/article-tigray-war-has-seen-up-to-half-a-million-dead-from-violence-and/ [8 Jul 2022]

[95] During the first meeting of high-ranking officials, China accused the U.S. human rights records https://www.wsj.com/articles/bitter-alaska-meeting-complicates-already-shaky-u-s-china-ties-11616185669?mod=article_inline and is issuing its own version of a human rights report about the U.S. https://www.newsweek.com/china-turns-tables-us-says-human-rights-violations-worsened-2021-1683478 [9 Jul 2022]

[96] Cain, Geoffrey (2021), The Perfect Police State: An Undercover Odyssey into China's Terrifying Surveillance Dystopia of the Future; https://www.amazon.com/Perfect-Police-State-Undercover-Surveillance-ebook/dp/B08HLP668T/ [7 Apr 2022]

[97] Mel Brooks in "History of the World: Part I" (1981), https://www.imdb.com/title/tt0082517/

[98] Fukuyama, Francis, The End of History and the Last Man (1992, and 2006), https://www.amazon.com/End-History-Last-Man/dp/0743284550

[99] https://www.intelligenteconomist.com/washington-consensus/ [9 Jul 2022]

[100] There are several slightly different versions of this quote but the underlying message was consistent. https://www.newsweek.com/two-days-before-christmas-george-washingtons-resignation-shocked-world-opinion-1478905 [9 Jul 2022]

[101] https://www.thebalance.com/trickle-down-economics-theory-effect-does-it-work-3305572 [9 Jul 2022]

[102] About the perception of economic power, see: https://www.pewresearch.org/global/2020/10/06/unfavorable-views-of-china-reach-historic-highs-in-many-countries/#in-europe-more-see-china-as-worlds-top-economic-power-than-u-s and about the general political system, see: https://www.theguardian.com/world/2021/may/05/us-threat-democracy-russia-china-global-poll

[103] https://www.macrotrends.net/countries/USA/united-states/life-expectancy

[104] https://hechingerreport.org/what-2018-pisa-international-rankings-tell-us-about-u-s-schools/ [9 Jul 2022]

[105] Philippon, Thomas (2019), The Great Reversal: How America Gave Up on Free Markets; https://www.amazon.com/Great-Reversal-America-Gave-Markets/dp/0674237544

[106] https://www.theguardian.com/world/2013/nov/05/new-zealand-law-email-phone-snooping [4 Aug 2022] and more recently also https://hardforking.com/new-zealand-passes-phone-access-law-you-can-be-asked-for-your-password/ [4 Aug 2022]

[107] It does not matter for the point made here that these school curricula are in most cases probably factually more correct than deviating opinions - but this means France is restricting people's *opinions*!

[108] In 2017, Japan was lightly ahead of the U.S. and China in the field of AI. In general, the U.S. is still ahead of China in the fields mentioned. Overall numbers of international patents seem dominated by China. https://stats.oecd.org/Index.aspx?DataSetCode=PATS_IPC and https://www.indexmundi.com/facts/indicators/IP.PAT.RESD/rankings [9 Jul 2022]

[109] U.S. Department of Energy (2021), Can the U.S. Compete in Basic Energy Science? Critical research frontiers and strategies; A report by the BESAC Subcommittee on International Benchmarking, 24 Aug 2021; https://science.osti.gov/-/media/bes/besac/pdf/202108/BESAC_Friend_202108.pdf?la=en&hash=9B261EE38257170AB25D0CE49AF73ECB62469C74 [7 Apr 2022]

[110] https://www.zdnet.com/article/u-s-vs-china-vs-india-in-engineering/ [7 Apr 2022]

[111] https://www.nextbigfuture.com/2017/08/future-tech-dominance-china-outnumber-usa-stem-grads-8-to-1-and-by-2030-15-to-1.html [7 Apr 2022]

[112] https://www.aip.org/fyi/2018/rapid-rise-china%E2%80%99s-stem-workforce-charted-national-science-board-report [7 Apr 2022]

[113] https://www.forbes.com/sites/niallmccarthy/2018/07/06/china-produces-more-cement-than-the-rest-of-the-world-combined-infographic/?sh=700e24536881 [7 Apr 2022]

[114] 2020 World Steel In Figures, World Steel Association (world steel.org)

[115] https://time.com/6090732/china-coal-power-plants-emissions/ [7 Apr 2022]

[116] For example smartphones.

[117] https://tradingeconomics.com/china/exports-of-goods-and-services-percent-of-gdp-wb-data.html [9 Jul 2022]

[118] See again the above-mentioned statistics published by the OECD (https://stats.oecd.org/Index.aspx?DataSetCode=PATS_IPC) and Index Mundi: https://www.indexmundi.com/facts/indicators/IP.PAT.RESD/rankings

[119] https://u.ae/en/about-the-uae/the-uae-government/government-of-future/innovation-in-the-uae [9 Jul 2022]

---

[120] For example, considering that 40% of all humans have not even reached half the threshold of energy use allowing them a dignified living, Vaclav Smil thinks that a lack of understanding of the underlying numbers make it extremely unlikely that the process could be anywhere near completion by 2050.
Smil, Vaclav, How the World Really Works: The Science Behind How We Got Here and Where We Are Going (2022)
https://www.amazon.com/How-World-Really-Works-Science-ebook/dp/B09CDB69WT

[121] Or at least logarithmically.

[122] Since nothing in nature can indefinitely grow exponentially, over time, this growth it more correctly: logarithmic. Eventually, it has to flatten out. For our purposes here, though, this will be many years in the future. We can therefore simplify the argument and ignore the eventual flattening out.

[123] For all those interested in deeper dives on the topic, visit https://singularityhub.com.

[124] https://www.weforum.org/whitepapers/our-shared-digital-future-responsible-digital-transformation-board-briefing-9ddf729993 [8 Jul 2022]

[125] Original quote: Thucydides (ca. 400 BC), The Peloponnesian War; See also: https://www.nku.edu/~weirk/ir/melian.html [7 Apr 2022]

[126] These is the number of overseas bases.
https://www.defensenews.com/opinion/commentary/2018/06/06/why-the-us-shouldnt-build-more-foreign-bases/ [9 Jul 2022]

[127] Michael O'Hanlon added up these numbers in chapter one of his 2021 book "The Art of War in an Age of Peace"
https://www.amazon.com/Art-War-Age-Peace-Restraint-ebook/dp/B093TMGC5C/ [13 Jul 2022]

[128] There are lots of debates about the topic. The bottom line seems to be that shooting several (some experts say 8) DF-17, for example, would have a very high likelihood to disable a carrier. https://airpowerasia.com/2020/05/08/aircraft-carriers-in-age-of-hypersonic-weapons/ [9 Jul 2022]

[129] Maybe even more than a tad. Hypersonic capabilities are nothing new to the U.S. The wake-up call is about us having to stop discounting Chinese technological prowess.

[130] The success of this approach can be questioned because the PRC *also* copies and therefore follows some of the U.S.' defense acquisition approaches. https://chinapower.csis.org/arms-companies/ [9 Jul 2022]

[131] https://news.usni.org/2021/03/08/panel-china-copied-silicon-valleys-model-to-outpace-the-pentagon-in-weapon-development and https://thediplomat.com/2019/07/innovation-in-the-new-era-of-chinese-military-power/ [9 Jul 2022]

[132] Brose, Christian (2020), The Kill Chain: Defending America in the Future of High-Tech Warfare; https://www.amazon.com/Kill-Chain-Defending-America-High-Tech-ebook/dp/B07W56RZCN/ref=tmm_kin_swatch_0?_encoding=UTF8&qid=1649364040&sr=1-5 [7 Apr 2022]

[133] https://basicresearch.defense.gov/

[134] https://www.diu.mil/about

[135] https://www.defensenews.com/opinion/commentary/2020/02/13/should-the-pentagon-stop-playing-the-role-of-venture-capitalist/ [7 Apr 2022]

[136] https://www.wsj.com/articles/send-in-the-marines-for-a-modernization-congress-plan-2030-forces-strikes-capability-11653940925

[137] https://www.aviationtoday.com/2018/09/27/harris-lm-icp/ [7 Apr 2022]

[138] https://www.inc.com/lisa-calhoun/mit-ranks-the-worlds-13-smartest-artificial-intelligence-companies.html [7 Apr 2022]

[139] https://www.forbes.com/sites/jimhandy/2013/03/25/moores-law-vs-wrights-law/?sh=2a0d3b4777d2

[140] https://history.army.mil/brochures/kw-chinter/chinter.htm [7 Apr 2022]

[141] https://nationalinterest.org/feature/the-five-biggest-disasters-american-military-history-11536 [7 Apr 2022]

[142] https://www.aei.org/op-eds/yes-taiwan-needs-nuclear-weapons-to-deter-china/ [9 Jul 2022]

[143] For more information see: https://www.belfercenter.org/thucydides-trap/overview-thucydides-trap [13 Jul 2022]

[144] Hugh White describes in detail Australia's geopolitical position in: White, Hugh, How to Defend Australia (2019) https://www.amazon.com/How-Defend-Australia-Hugh-White-ebook/dp/B07M8956X1/ [9 Jul 2022]

[145] Some of the main thinkers that keep pointing this out are Graham Allison at Harvard, John Mearsheimer at the University of Chicago, and Henry Kissinger. See also Kissinger's take on today's challenge, in: https://rielpolitik.com/2021/04/10/realpolitik-kissinger-accept-new-global-system-or-face-a-pre-wwi-geopolitical-situation/

[146] http://www.historytoday.com/archive/months-past/end-imperial-china [7 Apr 2022]

[147] https://www.hearing-history.co.uk/post/the-founding-of-the-nation-state-the-peace-of-westphalia [9 Jul 2022]

[148] Per American Asia Pacific Commander Philip Donaldson. https://www.theguardian.com/world/2021/mar/10/china-could-invade-taiwan-in-next-six-years-top-us-admiral-warns [7 Apr 2022]

[149] This is what Kevin Rudd mentions in his 2022 "Avoiding War."

[150] Many analysts with deep knowledge keep pointing out China's regional superiority in the Western Pacific, including Hugh White in an elaborate analysis (White, Hugh, The China Choice: Why We Should Share Power (2013), https://www.amazon.com/China-Choice-Should-Share-Power-ebook-dp-B00EDHTRBO/dp/B00EDHTRBO/), Dough Brandow of the Cato Institute (https://www.cato.org/commentary/pentagon-sizes-china-sees-serious-military-threat), and Henry Kissinger in: https://rielpolitik.com/2021/04/10/realpolitik-kissinger-accept-new-global-system-or-face-a-pre-wwi-geopolitical-situation/. [9 Jul 2022]

[151] Mihal, MajorChristopher (July-August 2021), Military Review, "Understanding the People's Liberation Army Rocket Force," July-August 2021, p. 16; https://www.armyupress.army.mil/Portals/7/military-review/Archives/English/JA-21/Mihal-PLA-Rocket-Force.pdf [7 Apr 2022]

[152] About the Taiwanese defense budget see https://www.globalsecurity.org/military/world/taiwan/budget.htm and about the training see: https://thehill.com/policy/international/579547-taiwan-increasing-training-of-reserve-forces/ [9 Jul 2022]

[153] See, for example: https://www.news.com.au/technology/innovation/military/uschina-war-over-taiwan-could-escalate-quickly-to-nuclear/news-story/113d892580aef6d4c0c1de682fbdee10 and

https://www.foreignaffairs.com/articles/china/2022-05-20/fight-over-taiwan-could-go-nuclear [9 Jul 2022]

[154] https://www.wsj.com/articles/what-real-economic-warfare-looked-like-russia-ukraine-world-trade-great-britain-11647637729 [7 Apr 2022]

[155] Leebaert, Derek, Magic and Mayhem: The Delusions of American Foreign Policy from Korea to Afghanistan (2011), https://www.amazon.com/Magic-Mayhem-Delusions-American-Afghanistan/dp/1439125716

[156] https://asiatimes.com/tag/foreign-policy-blob/ [9 Jul 2022]

[157] https://www.cnn.com/2022/02/01/politics/china-competition-bill-house/index.html [9 Jul 2022]

[158] https://alphahistory.com/worldwar1/treaty-of-versailles-military-restrictions-1919/ [7 Jul 2022]

[159] The following reference several of these war games but the number is much higher. https://asiatimes.com/2019/03/did-rand-get-it-right-in-its-war-game-exercise/ https://news.yahoo.com/were-going-to-lose-fast-us-air-force-held-a-war-game-that-started-with-a-chinese-biological-attack-170003936.html https://foreignpolicy.com/2022/04/03/taiwan-china-war-game-military-planning-strategy/ [7 Jul 2022]

[160] When we later talk about space-based weapons I will point out the destructive capabilities of kinetic energy in a more generic sense. For example, the dropping of rocks or steel rods from space can create such high velocities for a large enough mass that in certain situations they can come near to or even exceed the effect of traditional explosives.

[161] Dual-use refers to the use of a piece of equipment or technology for both civilian and military purposes.

[162] The formal goal is 2035, although this seems to indicate numbers of warheads, which is not a necessary condition. https://www.stripes.com/theaters/asia_pacific/2021-10-20/hypersonic-nuclear-China-missile-3306279.html [9 Jul 2022]

[163] https://www.airforcemag.com/chinas-nuclear-development-outstrips-predictions-1000-warheads-by-2030/ [9 Jul 2022]

[164] https://missilethreat.csis.org/country/china/ [7 Jul 2022]

[165] https://www.youtube.com/watch?v=ohubwZ64ySs and https://aps.org/policy/reports/popa-reports/missile-defense.cfm

[166] https://thehill.com/opinion/national-security/370737-our-enemies-could-use-nuclear-weapons-to-create-emp-attack/ [7 Jul 2022]

[167] These could be nuclear or non-nuclear EMPs, depending on the strategic goal.

[168] https://www.nytimes.com/2019/08/21/magazine/f35-joint-strike-fighter-program.html [9 Jul 2022]

[169] https://www.popularmechanics.com/military/navy-ships/a34965433/us-vs-china-navy-submarines/ [9 Jul 2022]

[170] https://www.forbes.com/sites/johncumbers/2020/02/03/china-now-out-invests-america-in-the-global-bioeconomy-by-30/?sh=5dc6f6b97440 [7 Jul 2022]

[171] https://www.cnn.com/2021/03/09/asia/russia-china-lunar-station-intl-hnk-scli-scn/index.html [9 Jul 2022]

[172] https://asiatimes.com/2021/11/helium-3-the-secret-mining-war-in-space/ [9 Jul 2022]

[173] Weichert, Brandon (2020), Winning Space: How America Remains a Superpower; https://www.amazon.com/Winning-Space-America-Remains-Superpower/dp/164572011X

[174] https://www.forbes.com/sites/ramseyfaragher/2020/08/01/chinas-homegrown-gps-is-now-fully-operational/?sh=33ef4a5d10f2 [7 Apr 2022]

[175] https://www.power-technology.com/features/feature-upgrading-us-grid-smart-self-healing-reality/ [7 Apr 2022]

[176] https://thediplomat.com/2021/10/the-battle-at-lake-changjin-and-chinas-new-view-of-war/ [7 Apr 2022]

[177] https://nationalinterest.org/blog/buzz/china's-j-20-stealth-fighter-now-equal-f-22-raptor-176117

[178] This is a bit of a wildcard because Taiwan itself, supported by the U.S., may be able to change this part of the equation. They may be able to establish a comparable challenge for Chinese forces during an attack on the island.

[179] https://www.surfpac.navy.mil/ccsg5/ and http://www.public.navy.mil/surfor/ccsg5/Pages/ourship.aspx [February 2022]

[180] https://www.19fortyfive.com/2021/12/df-21d-could-chinas-missiles-really-sink-a-us-navy-aircraft-carrier/

[181] https://www.19fortyfive.com/2021/11/df-21d-the-aircraft-carrier-killer-missile-that-makes-navy-admirals-freak-out/

[182] I use the term "non-kinetic" a bit loosely here, meaning the use of weapons that do not kill humans or blow up the buildings they live or work in.

[183] https://www.reuters.com/investigates/special-report/china-army-rockets/ [13 May 2022]

[184] Here are just two examples. On the American side, the launching of DF-21 missiles could look like a nuclear attack and trigger a disproportionate U.S. response. On the other side, China's still limited second strike options increases its likelihood to be the first one to pull the trigger and thus protect its first strike capabilities.

[185] Taylor, Brendan, The Four Flashpoints: How Asia Goes to War (2018) https://www.amazon.com/Four-Flashpoints-How-Asia-Goes/dp/1760640379 [9 Jul 2022]

[186] One can argue that China's fortifications in the South China Sea have already exactly done that.

[187] Like the Pratas Islands roughly halfway between Hainan and Taiwan. https://www.scmp.com/news/china/military/article/3137700/tiny-taiwan-island-could-be-trigger-us-china-clash

[188] https://www.history.com/this-day-in-history/united-states-announces-that-it-will-recognize-communist-china [5 Aug 2022]

[189] https://asiatimes.com/2021/07/how-far-would-japan-really-go-to-defend-taiwan/ [7 Jul 2022]

[190] https://www.world-nuclear.org/information-library/country-profiles/others/nuclear-power-in-taiwan.aspx [7 Jul 2022]

[191] https://nationalinterest.org/blog/buzz/chinas-greatest-nightmare-taiwan-armed-nuclear-weapons-80041 [7 Jul 2022]

[192] https://www.19fortyfive.com/2022/05/agile-combat-employment-how-the-us-air-force-would-fight-russia/ [13 Jul 2022]

[193] Drone-refueling could partially address this issue but without defenses, even drones can be shot down. https://www.barrons.com/news/us-navy-uses-drone-to-refuel-plane-during-flight-01623081614 [9 Jul 2022]

[194] https://www.forbes.com/sites/davidhambling/2021/04/30/us-navy-destroys-target-with-drone-swarm---and-sends-a-message-to-china/?sh=28b492f72df1 [7 Jul 2022]

[195] https://warontherocks.com/2019/02/drones-of-mass-destruction-drone-swarms-and-the-future-of-nuclear-chemical-and-biological-weapons/ [7 Jul 2022]

[196] https://www.martin-van-creveld.com/focus-on-taiwan/ [7 Apr 2022]

[197] Colby, Elbridge (2001), The Strategy of Denial: American Defense in the Age of Great Power Conflict; https://www.amazon.com/Strategy-Denial-American-Defense-

Conflict-ebook/dp/B09BMYDNPN/ref=tmm_kin_swatch_0?_encoding=UTF8&qid=1649360742&sr=8-1 [7 Apr 2022]

[198] Zero day attacks use vulnerabilities in computer code that have been found, or planted, but not communicated. Therefore, defenders could not develop fixes or countermeasures.

[199] https://blogs.cornell.edu/info2040/2018/09/17/tit-for-tat-approach-to-survival-evolutionary-strategies/ [9 Jul 2022]

[200] Particularly considering that in 2021 some Eastern European nations started to side with Taiwan, notably Lithuania.

[201] https://breakingdefense.com/2018/05/us-defense-budget-not-that-much-bigger-than-china-russia-gen-milley/ [9 Jul 2022]

[202] With the U.S. Civil War of the 1860s being an exception, at least to some degree. Then, though, the country was by far not as urbanized (according to the U.S. Census Bureau about 20% compared to today's 80%).

[203] About 249 million Americans lived in urban areas in 2010. https://www.census.gov/programs-surveys/geography/guidance/geo-areas/urban-rural/ua-facts.html [14 Apr 2022]

[204] High-voltage lines typically are at 345 kV, 500kV, or 765kV, industrial voltage is 220 or 440V, and residential and commercial voltage is 120 to 240V.

[205] https://www.cfr.org/backgrounder/how-does-us-power-grid-work [7 Apr 2022]

[206] Smil, Vaclav (2011), "Nitrogen cycle and world food production". World Agriculture. 2: 9-13.

[207] https://www.thoughtco.com/overview-of-the-haber-bosch-process-1434563 [7 Jul 2022]

[208] https://www.wired.com/2014/11/countdown-to-zero-day-stuxnet/ [6 Jul 2022]

[209] Pry, Peter (2021), Blackout Warfare: Attacking the U.S. Power Grid - A Revolution in Military Affairs. https://www.amazon.com/Blackout-Warfare-Attacking-Electric-Revolution-ebook/dp/B09DDB556M/

[210] https://spectrum.ieee.org/attack-on-nine-substations-could-take-down-us-grid

[211] https://www.popsci.com/scitech/article/2004-06/rods-god/

[212] https://www.reuters.com/article/us-pipeline-blast-alabama-idUSKBN12V2FC [13 May 2022]

[213] Per the U.S. Bureau of Transportation Statistics (BTS). https://www.bts.gov/content/us-oil-and-gas-pipeline-mileage [13 May 2022]

[214] https://www.visualcapitalist.com/submarine-cables/ [7 Apr 2022]

[215] https://www.submarinecablemap.com [7 Apr 2022]

[216] https://www.asiasentinel.com/p/taiwan-fears-china-cut-undersea-cables [7 Apr 2022]

[217] https://www.newsweek.com/undersea-cables-transport-99-percent-international-communications-319072 [7 Jul 2022]

[218] In both cases, the caveat is that some communication via satellite may still be possible.

[219] InterTubes: A Study of the US Long-haul Fiber-optic Infrastructure; http://conferences.sigcomm.org/sigcomm/2015/pdf/papers/p565.pdf [7 Apr 2022]

[220] Weichert, Brandon (2020), Winning Space: How America Remains a Superpower; https://www.amazon.com/Winning-Space-America-Remains-Superpower-ebook/dp/B08BTT14FH/

[221] https://www.euroconsult-ec.com/press-release/new-satellite-market-forecast-anticipates-1700-satellites-to-be-launched-on-

average-per-year-by-2030-as-new-entrants-and-incumbents-increase-their-investment-in-space/ [7 Jul 2022]

[222] https://www.energy.gov/sites/prod/files/2019/06/f64/Day%201%20-%20EPRI%20Electromagnetic%20Pulse%20-%20Horton.pdf [2 Aug 2022]

[223] https://www.theglobeandmail.com/politics/article-thirteen-canadians-detained-in-china-since-huawei-executives-arrest/ [7 Jul 2022]

[224] https://sampi.co/china-expat-population-statistics/ [7 Apr 2022]

[225] The numbers have grown throughout the past decade but the total number of immigrants in China is only just about 1 million, compared to over 10 million emigrants. https://www.migrationpolicy.org/article/china-development-transformed-migration [7 Jul 2022]

[226] https://www.wired.com/2016/10/inside-cyberattack-shocked-us-government/ [2 Aug 2022]

[227] Szu-chien Hsu, J. Michael Cole (2020), Insidious Power: How China Undermines Global Democracy; https://www.amazon.com/Insidious-Power-Undermines-Global-Democracy-ebook/dp/B08DWJCD4W

[228] https://www.wsj.com/articles/china-cracks-down-on-foreign-nonprofits-1425694223 [9 Jul 2022]

[229] https://www.amazon.com/gp/product/B000OZ0NXU/

[230] https://www.imdb.com/title/tt6189022/

[231] Over 30 years ago, in 1991, an early version of such a kinetic kill vehicle was successfully tested in the U.S. https://www.si.edu/object/vehicle-kinetic-kill-lightweight-exo-atmospheric-projectile:nasm_A20050460000

[232] https://www.bodyarmornews.com/how-does-kevlar-work/

[233] https://www.chemical-concepts.com/blog/2019/05/honeycomb-composites/

[234] David Ochmanek and Michael O'Hanlon are suggesting something similar in a practical scenario for defending Taiwan. https://thehill.com/opinion/national-security/584370-heres-the-strategy-to-prevent-china-from-taking-taiwan/ [9 Jul 2022]

[235] In simple terms, "on-the-loop" means observing and "in-the-loop" indicates a need for a human to actively participate in the process (e.g. by approving a specific action).

[236] https://www.techrepublic.com/article/google-employees-demand-end-to-companys-ai-work-with-defense-department/ [22 Jun 2022]

[237] Just one example of technology that already was available many years ago. https://phys.org/news/2013-04-agile-submarine-dash-milestones.html [8 jul 2022]

[238] https://www.doncio.navy.mil/ContentView.aspx?id=10869 [22 Jun 2022]

[239] https://nationalinterest.org/blog/reboot/nuclear-emp-attacks-are-threat-how-can-us-air-force-defend-itself-177612 [22 Jun 2022]

[240] An even more complete story would probably consider that Kodak kept pursuing an incorrect business model. Kodak survived but without being a serious contender in photography anymore. https://digital.hbs.edu/platform-rctom/submission/did-kodak-ignore-the-digital-revolution/

[241] https://www.forbes.com/sites/gregsatell/2014/09/05/a-look-back-at-why-blockbuster-really-failed-and-why-it-didnt-have-to/ [9 Jul 2022]

[242] Hill, Napoleon (1937), Think and Grow Rich; https://www.amazon.com/Think-Grow-Rich-Original-Classic-ebook/dp/B07N576VN1/

[243] One way to start going about this would be to look at the share of costs if one takes out those for raw materials, capital investment,

and manual labor, and do so also for the supply chains of equipment and semifinished products. As a result, the costs of most complex products would consists mostly of "digitized" inputs, meaning they could be performed from anywhere and by anyone suitably trained and digitally connected.

[244] Webb, Amy (2019), The Big Nine: How the Tech Titans and Their Thinking MachinesCould Warp Humanity; https://www.amazon.com/Big-Nine-Thinking-Machines-Humanity-ebook/dp/B07H7G7CMN/

[245] Obviously, the degree and the manner of cooperation is different in the U.S., compared to China. But particularly Microsoft, Amazon, Google, and IBM are closely cooperating with the U.S. government. Facebook, Google, and Amazon, even Apple, all take their clues about freedom of expression - from the government.

[246] https://www.defense.gov/About/ [1 Aug 2022]

[247] https://irp.fas.org/budget/ [21 Jun 2022]

[248] https://www.cnbc.com/2022/03/05/china-defense-spending-to-rise-by-7point1percent-in-2022-says-finance-ministry.html

[249] https://www.nber.org/system/files/chapters/c1567/c1567.pdf

[250] https://ourworldindata.org/employment-in-agriculture [22 Jun 2022]

[251] https://fred.stlouisfed.org/series/USAPEFANA [22 Jun 2022]

[252] https://www.nam.org/state-manufacturing-data/2020-united-states-manufacturing-facts/ [22 Jun 2022]

[253] However, where governments do not have the right to track and surveil people, businesses in the West do have this right. Governments avail themselves to their information and either pay for this, or at a minimum protect Big Data corporations from being limited in their ability to conduct total surveillance and amass the

most intimate data about each person. See the next footnote about Shoshana Zuboff's book, which is only one of many.

[254] Zuboff, Shoshana (2019), The Age of Surveillance Capitalism: The Fight for a Human Future at the New Frontier of Power; https://www.amazon.com/Age-Surveillance-Capitalism-Future-Frontier-ebook/dp/B01N2QEZE2/

[255] https://www.forbes.com/sites/jamesbroughel/2022/04/21/cascading-risks-the-us-is-vulnerable-to-an-electromagnetic-pulse-attack/ [22 Jun 2022]

[256] The first such solution has been announced by a key U.S. government agency in July 2022. https://www.nist.gov/news-events/news/2022/07/nist-announces-first-four-quantum-resistant-cryptographic-algorithms [1 Aug 2022]

[257] https://www.defenseone.com/technology/2020/08/heres-theme-powering-armys-new-communications-tech/167681/ [12 Jun 2022]

[258] https://phys.org/news/2021-01-world-quantum-network.html

[259] https://worldfinancialreview.com/how-quantum-technology-is-used-in-finance/ [11 Jul 2022]

[260] https://www.digitaltrends.com/space/space-based-solar-power-plants-could-soon-be-a-thing/ [11 Jul 2022]

[261] https://cleantechnica.com/2018/07/01/peeking-behind-teslas-labor-curtain/

[262] https://www.wsj.com/articles/singapore-manufacturing-factory-automation-11655488002?page=1 [22 Jun 2022]

[263] https://3dinsider.com/us-navy-turning-aircraft-carriers-into-3d-printing-factories/ [22 Jun 2022]

[264] https://www.bbc.com/news/magazine-16870579 [7 Apr 2022] This number keeps being mentioned, but since the publication of this article the world population increase by almost another one billion people.

---

[265] Like how to deal with the effect of cosmic radiation and the lack of an atmosphere on blood and health.

[266] https://www.huffpost.com/entry/10-lessons-the-us-can-lea_b_1143462 [11 Jul 2022]

[267] See https://www.politico.com/news/2020/11/04/china-us-economy-industrial-policy-global-translation-433954 and https://www.discoursemagazine.com/economics/2021/03/11/should-the-u-s-copy-chinas-industrial-policy/ [11 Jul 2022]

[268] https://thoughteconomics.com/peter-diamandis/ [11 Jul 2022]

[269] See his two most recent books: Mahbubani, Kishore, The Asian 21st Century (2022) https://www.amazon.com/Asian-21st-Century-China-Globalization-ebook-dp-B09PJCLZSL/dp/B09PJCLZSL/ [11 Jul 2022]

[270] de la Cruz, Diego Sánchez (2017), Por qué soy liberal; https://www.amazon.com/Por-qu%C3%A9-soy-liberal-liberalismo-ebook/dp/B01N6V6P69/
This is from the Epilogue of the book.

[271] https://www.kurzweilai.net/the-law-of-accelerating-returns [10 Jun 2022]

[272] https://www.brookings.edu/research/a-dozen-facts-about-the-economics-of-the-u-s-health-care-system/

[273] Silverglate, Harvey; Dershowitz, Alan (2011), Three Felonies a Day: How the Feds Target the Innocent; https://www.amazon.com/Three-Felonies-Day-Target-Innocent/dp/1594035229

[274] O'Hanlon, Michael (2021), The Art of War in an Age of Peace: U.S. Grand Strategy and Resolute Restraint; https://www.amazon.com/Art-War-Age-Peace-Restraint-ebook/dp/B093TMGC5C/

[275] https://www.hks.harvard.edu/sites/default/files/HKSEE/HKSEE%20PDFs/Allison_Spheres%20of%20Influence%2C%20Foreign%20Affairs%2C%20March-April%202020.pdf

[276] https://foreignpolicyblogs.com/2022/01/21/taiwan-and-ukraine-beyond-great-power-competition/

[277] White, Hugh (2012), The China Choice - Why America Should Share Power; https://www.amazon.com/China-Choice-Should-Share-Power-ebook/dp/B00EDHTRBO/

[278] This is something that China has done *in principle*, but refused to do when specific issues were at stake, like the ownership over islands.

Index

D

E

F

G

H

I

J

K

L

M

N

O

P

Q

R

S

T

U

W

X

Z

Made in the USA
Middletown, DE
11 October 2022